For a complete listing of titles in the
Artech House Power Engineering Series,
turn to the back of this book.

Renewable Energy Technologies and Resources

Nader Anani

ARTECH
HOUSE

BOSTON | LONDON
artechhouse.com

Library of Congress Cataloging-in-Publication Data
A catalog record for this book is available from the U.S. Library of Congress.

British Library Cataloguing in Publication Data
A catalogue record for this book is available from the British Library.

Cover design by John Gomes

ISBN 13: 978-1-63081-573-8

The solutions manual for this book is available at www.artechhouse.com.

© 2020 ARTECH HOUSE
685 Canton Street
Norwood, MA 02062

10 9 8 7 6 5 4 3 2 1

Contents

1

Introduction

Aim: Define the concepts and terminology related to energy.

1.1 Learning Outcomes

After actively engaging with the material in this chapter, you should be able to

1. Distinguish between a renewable energy resource and a renewable energy technology.
2. Explain the basic forms of energy;
3. Define the terms primary, delivered, and useful energy;
4. Discuss the adverse effects of fossil fuels on the environment;
5. State the main sources of renewable energy;
6. Calculate the capacity factor of a power plant.

1.2 Overview

This chapter presents a simplified review of basic physical concepts related to the world of energy for the benefit of readers who may not have a scientific background. It defines physical quantities, such as energy, force, and power, and their units and multipliers. The chapter explains the basic forms of energy: kinetic, gravitational, electrical, and nuclear energy. Moreover, the chapter defines and explains the significance of the capacity factor of a power plant in determining its productivity. In addition, the chapter discusses the adverse effects of using fossil fuels on the environment and the need to switch to renewables. Finally, the chapter presents an outline of different renewable energy sources and their estimated global resources.

1.3   Introduction

Resources of energy are classified as either renewable or nonrenewable (also known as conventional) energy resources. Conventional energy is derived from sources, such as coal, oil, and gasses, that are exhaustible. Renewable energy, on the other hand, is derived from resources, such as wind, solar, and biomass, that are sustainable or replenishable. Humans have been using the energy in sunlight and kinetic energy in wind for thousands of years and both (i.e., sunlight and wind energy) are still abundant. When we plant a tree and burn it for energy, we can plant another one and in 5 to 10 years or so, we have another tree to burn (i.e., biomass can be made a replenishable source of energy). However, a tree that has turned into coal took millions of years to form; hence coal is not replenishable. Renewable energy has been defined as the energy obtained from the continuous or repetitive currents of energy recurring in the natural environment [1]. For most renewable energy sources, the fundamental source of energy is derived essentially from the sun. Therefore, most renewable energy sources are, in principle, solar sources. Sun energy heats the earth and oceans, which results in the movement of waves and wind. Waves and wind energy are, therefore, fundamentally manifestations of solar energy. Solar energy may be used directly, for example, for heating and lighting, and indirectly solar energy can be used for heating water to produce steam to turn turbines to generate electricity.

Conventional sources of energy based on fossil fuels, such as coal, oil, and gas, have been humanity's main source of energy for more than a century. They are still the predominant source of energy and it is estimated that they accounted for nearly 80% of the total global energy consumption in 2015 [2]. However, the dependence on fossil fuels has negative impacts on the environment because they produce harmful gases, such as carbon dioxide (CO_2), which are responsible for adverse effects, such as climate change. In addition, fossil fuel supplies are neither secure nor sustainable and are subject to political instability. These factors led governments around the world to search for alternative sources of energy that are renewable or sustainable. The main features of a renewable energy source are as follows: it is not significantly depleted with continual use and its harvesting and use does not result in any significant adverse effects on the environment or public health, and does not result in any form of social injustice [3].

At this point, it is worth explaining the difference between a renewable energy resource and a renewable energy technology. For example, wind, tidal, wave, and sunlight are resources of energy, while a wind turbine or a photovoltaic panel are examples of renewable energy technologies. Therefore, renewable energy technology refers to devices and systems that are used to harvest renewable energy and convert it to a useful form ready for use as a fuel.

1.4 Energy, Work, and Power

We need energy to do work (for example, for heating or moving objects, providing light, cutting objects, and generating sound). Energy is stored or can be stored in various forms and although we say we consumed or spent energy in, say, moving an object from one place to another, energy can neither be consumed nor be created from nothing. Energy can only be converted from one form to another resulting in work done, or energy transfer, as a result of this conversion process. If an object is stationary or moving at constant velocity, then you need force to move it or alter its direction of motion and/or its speed. The force (F) is measured using the International System (SI) unit newton (N), which is defined as the magnitude of force required to give an object of mass (m) of 1 kg an acceleration (a) of one meter per second. That is, force is given by

$$F = m \times a \qquad (1.1)$$

When you apply a force F (N), you do work or convert energy. The work done in joules, W (J), in moving an object a distance d (m) is given as

$$W = F \times d \qquad (1.2)$$

Therefore, the joule (J), is defined as the energy delivered by a force of one newton causing a movement through one meter. The work done is the energy transferred or converted in moving the object. Hence, we will use the terms work done and energy interchangeably. In the above it is assumed that the line of action of the force is parallel to the direction of motion. If the force F is at an angle, say $\theta°$, with respect to the direction of motion, then the work done is the product of the distance moved and the component of the force parallel to the direction of motion

$$W = Fd \cos \theta \qquad (1.3)$$

If it takes you, say, t seconds to move the object, the energy converted (or transferred) per second defines the power P in watts (W) as

$$P = \frac{W}{t} \qquad (1.4)$$

That is, power is the rate at which energy is being converted from one form to another, and in general, power is expressed as

$$P = \frac{dW}{dt} \qquad (1.5)$$

Therefore, one watt is equivalent to one joule of energy converted or transferred per second.

As an example, when a 100W lightbulb is turned on 10 minutes, it converts an amount of electrical energy equals to

$$W = 10 \, (\text{min}) \times 60 \, (\text{s}) \times 100 \, (\text{W}) = 60000 \ \text{J}$$

This energy is converted to light and waste heat. Consider as another example, a 1-kW heater switched on for one hour; the amount of electrical energy it uses is

$$\begin{aligned} W &= P \times t \\ &= 1000 \, (\text{W}) \times 60 \times 60 \\ &= 3.6 \times 10^{6} \\ &= 3.6 \ \text{MJ} \ \ (\text{i.e. 3.6 million joules}) \end{aligned}$$

This is a very large number; therefore, using the joule as a unit for measuring domestic energy consumption is impractical. Hence, an alternative unit of energy is used that measures the energy use over a specific period, usually an hour. Indeed, the basic unit adopted for this purpose is the kilowatt-hour (kWh), which uses kW instead of watt and hours instead of seconds. That is 1kWh = 3.6 MJ. Multipliers and their prefixes of energy units are shown in Table 1.1. Energy may also be also measured in terms of quantities of fuel, for example, using units such as tonnes of coal equivalent (TCE), tonnes of oil equivalent (TOE), and barrels of oil equivalent (BOE).

Another unit, typically used for energy in food, is the calorie. One calorie is the amount of energy required to raise 1 kg of water by one degree Celsius. To give you an idea of the relative sizes of joules and calories, one calorie is equal to 4,184 joules. An average adult needs to burn 2,000 Calories in a day to maintain daily health and life, which amounts to approximately 8.4 million joules.

Table 1.1
Multiplier Prefixes

Symbol	Prefix	Example	Multiply By	As Power of Ten
K	Kilo-	Kilowatt	One thousand	10^3
M	Mega-	Megawatt	One million	10^6
G	Giga-	Gigavolt	One billion	10^9
T	Tera-	Terawatt	One trillion	10^{12}
P	Peta-	Petawatt	One quadrillion	10^{15}
E	Exa-	Exajoule	One quintillion	10^{18}

1.5 Basic Forms of Energy

There are several renewable energy technologies that convert one form of energy into another and in most cases, the final form of energy is electrical energy. In any such conversion of energy, the total quantity of energy remains constant. This principle—that energy is always conserved—is expressed by the first law of thermodynamics [4]. So, if the electrical energy output of a power plant is less than the energy content of the fuel input, then some of the energy must have been converted to another form, which is usually waste heat, also known as stray energy. Hence, although in everyday language we speak of consuming energy, we don't. When we burn fuel in a car engine, we simply convert the stored chemical energy in the fuel into heat and kinetic energy that moves the car. Similarly, when we use a wind turbine, we extract kinetic energy from the moving wind and convert it into electrical energy. In any such conversion, there will always be some energy lost; hence the efficiency of any energy conversion system is always less than 100%. In general, we categorize energy into four basic forms: kinetic energy, electrical energy, nuclear energy, and gravitational energy. This is because any other form of energy can be referred to one of these four basic forms.

1.5.1 Kinetic Energy

An object of mass m (kg) moving with a velocity of v (*m/s*) possesses an amount of kinetic energy (E) in joules (J) given by

$$E = \frac{1}{2}mv^2 \tag{1.6}$$

At atomic level, the kinetic theory of matter states that all matter is made up of small particles—atoms or molecules—that are continuously moving or vibrating and colliding with each other [4]. This continuous motion of atoms or molecules determine the temperature of the substance. If heat is added to a substance, the atoms or molecules vibrate faster. Therefore, heat energy may be considered a manifestation of kinetic energy. Temperature is measured in kelvin (K) and 0K corresponds to zero molecular motion. Temperature can also be measured using the Celsius (°C) scale. The size of one degree on both scales is the same. The 0 degree Celsius corresponds to the freezing point of water while 100 degrees Celsius is the boiling point of pure water at one atmospheric pressure. The two scales are related by

$$\text{Temperature in K} = \text{temperature in } °C + 273 \tag{1.7}$$

1.5.2 Gravitational Energy

The earth possesses a gravitational field that pulls objects placed in this field toward the center of the earth. Therefore, it requires work to lift an object of mass m (kg) from the surface of the earth to a height h (m) above the surface. In the new position above the earth, the object possesses a gravitational potential energy due to its position above the earth.

The magnitude of the force in newtons (N) required to lift an object of mass m (kg) is called the weight (W) defined as

$$W = mg \tag{1.8}$$

Where g (m/s^2) is the constant acceleration due to gravity and is approximately 9.8 m/s^2.

The gravitational potential energy, or simply the potential energy P (J), of the object at a height h (m) above ground is given by

$$P = mgh \tag{1.9}$$

1.5.3 Electrical Energy

At atomic scale, electrical forces hold together the atoms and molecules of all substances. Every atom has a cloud of electrically charged electrons, moving continually around a central nucleus. During a chemical reaction, atoms bond with other atoms to form molecules, and this leads to changing the distribution of electrons. Therefore, chemical energy, at atomic level, may be seen as a form of electrical energy.

At a larger scale, the scale we are familiar with, electrical energy is carried by the steady flow of free electrons in conductors, which we call electric current.

In conductors of electricity, like copper, some electrons can become detached from their parent atoms and are able to move freely within the lattice structure of the conductor. However, these free electrons move in random fashion and therefore don't contribute to any net flow of current. To make these electrons flow in one direction resulting in steady flow of current, an external source of energy, such as a battery or a generator, is required. Consider the simple electric circuit shown in Figure 1.1

The source of energy is the DC voltage source that provides an electro-motive force (EMF) of E volts (V). This EMF introduces a potential difference *V* measured in volts (V) across the load R and circulates a steady current *I* measured in ampere (A) around the circuit. Note that the direction of the conventional current flow is opposite to the direction of the flow of electrons. By convention, current leaves the positive side of a voltage sources and goes into the positive side of a load. The load current *I* (A) is proportional to the voltage difference, or voltage drop, *V* (V) across the load and is given by Ohm's law as

$$I = \frac{V}{R} \tag{1.10}$$

Where *R* is the electrical resistance of the load and is measured in ohms (Ω). An electron carries a negative charge of -1.6×10^{-19} coulomb (C) and a conductor is said to have a current of one ampere when charge flows though it at a rate of one coulomb per second. When these electrons flow through an electrical load, the load converts the electrical energy into another form of energy. For example, if the load is a heater, the electrical energy is converted to heat and if it is a lightbulb, it is converted to light. Note that, for an ideal voltage source (i.e., one that has no internal resistance) the EMF E is numerically equal to the voltage drop, V, across the load resistance (i.e., $E = IR$).

The power in watts (W) delivered by the electrical energy source to the load is given by

$$\text{power (W)} = V \times I \tag{1.11}$$

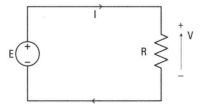

Figure 1.1 A simple electric circuit with a load resistance R and a battery E (V).

If the electrical source is a periodic function of time, then V and I in the above equation are the root-mean-square (RMS) values. For example, if the voltage is a sinewave with maximum amplitude V_m (V), then the RMS value, also known as the effective value, is calculated as

$$V = \frac{V_m}{\sqrt{2}} \tag{1.12}$$

An electrical load or an appliance is normally rated by its operating voltage and power. For example, if a resistive heater is rated as 230V 2 kW, this means that the RMS value of the input voltage is 230V and the RMS current in ampere (A) it draws from the supply is 2000 $W/230$ $V = 8.7$ A. The energy it consumes depends on how long it is switched on; for example, if the heater is used for 2 hours, the energy it consumes, or to be more precise, the electrical energy it converts, is

$$W = (2 \text{ kW}) \times (2 \text{ hours})$$
$$= 4 \text{ kWh}$$

An electricity meter in a building measures the number of kWh used in the building and the cost of electrical energy is therefore calculated as

$$\text{Cost of Energy} = \text{Units used (kWh)} \times \text{cost per unit (cents)} \tag{1.13}$$

1.5.4 Nuclear Energy

Nuclear energy is contained in the center of atoms (i.e., in the nuclei), which when released provides huge amounts of energy. There are two methods for freeing this energy: nuclear fission and nuclear fusion. In nuclear fission, the nucleus of a heavy atom is split into smaller nuclei inside a nuclear reactor. For example, when the nucleus of uranium-235 is bombarded with a neutron it splits into two smaller nuclei known as uranium 236, which is highly unstable. This unstable nucleus is then split into two nuclei with the release of huge amounts of energy and the emission of two or three more neutrons. These released neutrons can instigate further fissions by interacting with fresh nuclei of uranium-235that will then emit new neutrons and so forth. This multiplying effect is called a chain reaction and a huge amount of energy is released, which is used to boil water to produce steam to drive turbines that turn electric generators.

In nuclear fusion, two light nuclei join to make one heavy nucleus. This is the kind of reaction that occurs in the sun, where two hydrogen nuclei fuse together under extremely high temperature and pressure forming the nucleus of a helium isotope. The main problem with fusion is the fact that it requires the fusing of two nuclei that have the same type of charge (i.e., positive charge), which means that they will repel each other. Therefore, the fusion must happen fast in order to prevent the repulsive forces taking place. One way to force particles to travel fast is to place them in a hot gas or in plasma. This requires extremely high temperatures and the technology to achieve this is still under development.

Nuclear power plants generate electricity with relatively very low emissions of greenhouse gasses and hence, their contribution to the global climate change is relatively small. In addition, nuclear energy technology is well-established and therefore reliable. Further, a nuclear power plant can provide electricity almost 24/7. However, there are some proclaimed disadvantages of nuclear power generation: waste is radioactive and is difficult and expensive to store safely. The fuel for nuclear energy is uranium, which is a scarce and exhaustible resource. The capital cost of a nuclear power plant is relatively expensive and its construction time is considerably long. Finally, there is always the risk of an accident, which can be catastrophic as it is impossible to build a 100% accident-proof plant.

1.6 Capacity Factor and Efficiency

A power plant cannot run constantly all the time without interruption (i.e., cannot be available for power generation 24/7), which limits its energy productivity. For example, a plant may have to be shut down for regular services or repairs. Renewable energy plants have an added limitation on their availability and energy productivity because their output energy is dependent on weather conditions. For example, a wind turbine cannot generate energy if there is no wind and similarly, a photovoltaic plant cannot produce energy in the absence of the sun. On average, a wind turbine typically runs for 2,000 to 4,000 hours per year depending on its location. To account for the productivity of a power plant, the capacity factor (CF), also known as the load factor, is defined. The CF is a figure of merit for any electricity generating plant and it must be taken into consideration when assessing the economics of any power plant. The CF of a plant is defined as the actual output energy over a given period, typically a year, divided by the maximum possible output over the same period.

$$\text{CF} = \frac{\text{Actual output energy per year}}{\text{Maximum possible output energy per year}} \qquad (1.14)$$

There are 24(hours) × 365(days) = 8760 hours in a year, and therefore, the maximum possible energy output, over a year, is given by

$$\text{Maximum possible output} = 8760(\text{hours}) \times \text{Rated Capacity} \qquad (1.15)$$

The rated capacity, also known as the installed capacity, is the maximum power a plant has been designed to supply, typically measured in MW or GW. The CF can be expressed as either decimal fraction or a percentage. For example, consider a power plant rated at 10 MW. If the plant runs throughout the year without interruptions, it will generate the maximum possible output (i.e., 8760h × 10 MW = 87600 MWh) and its capacity factor will be 100% or 1. On the other hand, a 10-MW wind turbine cannot produce electricity all the time simply because wind is not continuously available, and when it is available it is not always blowing at the full-rated speed of the turbine to allow it to produce its full rated capacity. Typically, a 10-MW wind turbine generates about 25,000 MWh. Therefore, its capacity factor would be

$$\frac{25000}{87600} \times 100\% = 28.5\% \text{ or } 0.285$$

When energy is converted from one form to another, the useful energy output is always less than the input. The efficiency of conversion is defined as the ratio of the output to the input energy and is usually expressed as a percentage

$$\text{efficiency} = \frac{\text{energy output}}{\text{energy input}} \times 100\% \qquad (1.16)$$

Typical efficiencies are around 35%–40% in a coal-fired power plant, around 26% for a wind power plant, and as low as 12% for a solar photovoltaic plant [5]. However, these figures are for guidance only and they can vary considerably depending on factors such as the technology and the climate of the location. Although in many situations, efficiency can be improved by better design, in many others the efficiency is limited by the nature of energy conversion, which the designer has no control over. The principle is that whenever we convert energy from a low-quality form, such as heat, into a high-quality form (electrical or mechanical energy) there is always a physical or natural upper limit on efficiency. Heat is due to randomly moving molecules, which is essentially a chaotic or low-quality form of energy. Electrical energy is due to the ordered flow of charges and is a high-quality form of energy. The kinetic energy in the wind is also chaotic (i.e., low-quality energy) and therefore, when this energy is converted to a high-quality rotational mechanical energy using a wind turbine, the maximum theoretical conversion efficiency is limited to about 59% [6]. Another example of efficiency limitation is in a photovoltaic system. A perfectly designed silicon-based photovoltaic panel (i.e., one with no losses at all) cannot achieve an efficiency of higher than about 49% [7]. This inherent limit on efficiency is due to the distribution of energy in the spectrum of the sunlight arriving on Earth and the bandgap energy of silicon.

1.7 Current Use of Energy

In a coal-fired power station, coal is burned to generate steam that turns a turbine, which turns a generator to generate electricity. The electricity is then transmitted using wires to households. Losses are incurred in each conversion process and during the transmission. The energy released at the power station when the coal is burned is called the primary energy required for that use. The amount of electricity reaching the consumer, after conversion and transmission losses, is called the delivered energy. If the delivered energy is, say, used to heat water in a tank, further losses are incurred in the tank and pipes, and the final quantity that comes out of the hot water tap is called the useful energy. The world's total annual consumption of all forms of primary energy increased more than tenfold during the twentieth century, and by the year 2017 had reached an estimated 560 EJ (exajoules), or some 14,000 million TOE. As shown in Figure 1.2, fossil fuels provided more than 80% of the total. This represented a growth of 2.2% compared with the previous 10-year average of 1.7%. Further, carbon emissions increased by 1.6% after almost no growth from 2014 to 2016 [8]. Figure 1.2 shows the contributions of the renewable energy sources, excluding hydropower, to the total world's primary energy supplies. When hydropower is included, renewable sources contributed to an estimated 25% of world primary

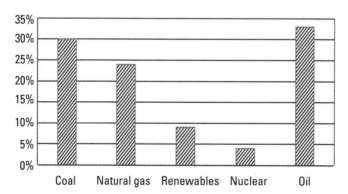

Figure 1.2 Percentage contributions to world primary energy consumption in 2017. Renewables exclude hydropower.

energy in 2017 [9]. The largest contribution is an estimated 30 EJ from traditional biomass (wood, straw, dung, etc.) mainly used in developing countries. Since most of this isn't traded, it doesn't often enter national economic statistics and its true magnitude is only known approximately [3]. The next largest category is new biomasses, which includes wood and other crops specifically grown for energy for producing fuels (e.g., biogas and biodiesel).

At the current rate of consumption, it is estimated that world coal reserves could last for about 120 years, oil for approximately 45 years, and natural gas for around 60 years [10]. However, in the more immediate future there are likely to be serious constraints on the rate at which fossil fuels can be produced, particularly oil. Statistics indicate, for example, that global production of oil is coming up to its limit and for every new barrel of oil found, four barrels are consumed. Current oilfields have a limited life; once exhausted they must be replaced with new ones. In order to just maintain the world's oil production at its current level, many new oil fields will have to be developed. Even more challenging is the need for new fields to be continuously discovered. According to the International Energy Agency, the easy oil has been largely used up. What remains is likely to be more expensive and is located in difficult areas such as the Arctic or in deep offshore wells [11].

1.8 Fossil Fuels and Greenhouse Effect

Our climate is changing; the ice is melting at the north and south poles, seawater levels are rising, the average temperature is generally rising, our energy resources are depleting, and pollution and diseases are also on the rise. The prime reason for these changes is the greenhouse effect. So, what is the greenhouse effect? Well, the greenhouse effect is actually a good thing! It is a natural phenomenon that allows our planet Earth to be habitable. Without the greenhouse effect the average temperature on Earth would be −18 degrees Celsius. In a greenhouse, the visible light from the sun penetrates the glass panes and is partly absorbed by the plants and the soil and partly reflected again into the glass panes, inside the greenhouse, as invisible heat radiation. Some of this invisible radiation leaves the greenhouse and some is sent back into the greenhouse, which keeps the greenhouse warm and allows us to grow plants in cooler locations. That is the glass panes in a greenhouse provide a shield from the cold weather outside and traps heat to keep the greenhouse warm. This greenhouse phenomenon occurs also naturally for our planet Earth. The atmosphere around the earth does the same job as the glass panes in a greenhouse. Some of the light from the sun penetrates the atmosphere and is partly absorbed by the earth and partly bounced again into the space as invisible radiation. Some of this invisible radiation leaves atmosphere and some is reflected, by the atmosphere, back onto

Earth which keeps the earth at an average temperature of about 15 degrees Celsius. If the atmosphere did not trap some of this radiation, the temperature on Earth would be about −18 degrees Celsius which would render the earth inhabitable. The atmosphere is composed of natural gases, of which nitrogen and oxygen constitute over 98%. In addition, there are other gases, such as carbon dioxide (CO_2), methane (CH_4), nitrous oxide (N_2O), and water vapor (H_2O). Artificial gases called chlorofluorocarbons (CFCs) are also present in the atmosphere. These are collectively known as greenhouse gases and when they are present in the correct proportion, they act like the shield around the earth protecting it from harmful radiation and trapping the right amount of heat energy to keep its temperature at levels that keeps it warm enough to harbor life. However, when the concentration of these greenhouse gases increases above the right level, by human activities, they trap more heat than necessary causing rise in temperature. This effect known as global warming has many adverse effects on the environment: it causes ice to melt, a rise in sea levels, higher rate of water evaporation that will eventually cause severe weather conditions, spread of diseases, and desertification. A major contributor to the global climate change is the emission of carbon dioxide from the burning of fossil fuels in power generation and internal combustion engines. This led to a growing consensus for the need to reduce such emissions. In order to ensure that global mean temperature rises do not exceed 2°C above preindustrial levels by 2050, studies show that global carbon emissions must be reduced by approximately 80% by that date [3]. Emission reductions on this scale will inevitably require global cooperative and drastic actions, such as a switch to low- or zero-carbon energy sources (i.e., renewables).

1.9 Renewable Energy Sources

The principle source of most renewable energy sources is solar radiation. Solar radiation arriving at the surface of the earth amounts to 5.4 million EJ per year, of which about 30% is reflected into space. The remaining 3.8 million EJ per year is available for utilization on Earth [3]. This is approximately 10,000 times more than the rate of fossil and nuclear fuel consumption in 2015 [12]. It is the sun that heats the oceans and the earth, thus producing waves and wind. Hence, waves and wind are solar sources. Two nonsolar sources are tidal and geothermal sources. Tidal energy is mainly due to the movement of the moon, while geothermal energy is due to heat from the interior of Earth.

1.9.1 Solar Energy

Solar energy is of two types: solar thermal and solar electric or photovoltaics. Photovoltaic (PV) systems use PV modules that convert energy in the sunlight directly into electricity. In 2017, the global PV generation reached a capacity of about 2% of the global power output at around 400 GW and is expected to increase to nearly 600 GW by 2022 [13]. The drawbacks of PV generation include output intermittency due to varying weather conditions, relatively low efficiency of PV modules, low efficiency that requires large areas for installation, and it is still relatively expensive. Solar thermal technologies use special devices to collect and direct the sun's rays to heat water in special boilers to produce steam that can be used to turn electrical generators.

1.9.2   Hydropower

The energy in moving (or falling) water is used to turn turbine-generator set to convert this energy into electricity. This is considered a renewable energy because the water cycle is constantly renewed by the sun and hence is inexhaustible. Hydropower makes around 70% of the global renewable source for generation of electricity. Its main advantages are its predictability, consistency, and flexibility, which makes it capable of meeting base and peak load demands [14].

1.9.3   Wind Power

Moving wind possesses kinetic energy that depends on its speed. This kinetic energy can be deployed to rotate the blades of a wind turbine coupled to an electric generator to produce electricity. In 2017, the installed capacity of wind power, excluding hydropower, added 29% of the global installed capacity of renewable energy sources [15].

1.9.4   Wave Power

Where winds blow over long stretches of ocean, they create waves, and a variety of technologies can be used to extract that energy. One such technology is to force a wave to push water into a chamber causing air to rise and fall as it leaves. The rise and fall of air can then be used to turn a turbine and generate electricity. Wave energy is abundant and predictable, therefore allowing calculation of the amount of electricity that can be generated. However, it requires the installation of large machines in the water that can upset the marine life [16].

1.9.5   Biomass

Biomass refers to all plants and animal materials that can be burned or digested to release energy. Examples of biomass materials are wood, straw, and animal waste. Some plants like sugarcanes and corn are grown specifically to be used in the production of biofuels, which can be used in both petrol and diesel engines to replace oil. Estimates of this resource vary considerably; some put it at about 100 EJ (i.e., about 20% of the current global demand of primary energy) [17].

1.9.6 Tidal Energy

Ocean tides are caused by the gravitational pull of the moon (with a small contribution from the sun) on the world's oceans, causing a regular rise and fall in water levels as the earth rotates [18]. The power of the tides can be harnessed by building a low dam or barrage, behind which the rising waters can be captured and then allowed to flow back through electricity-generating turbines. Advantages of tidal energy include their relatively long lifespan, which reduces the cost of the generated electricity, and the predictability of the generation since the tidal currents are highly predictable. The proclaimed disadvantages of tidal power generation include the relatively high cost of construction and their impact on marine life.

1.9.7 Geothermal

Heat from within the earth is the source of geothermal energy. The high temperature of the earth's interior was originally caused by gravitational contraction of the planet as it was formed but has since been enhanced by the heat from the decay of radioactive materials deep within the earth. The sources of this heat energy range from shallow ground to extremely hot and molten rocks several kilometers below the earth's surface [19]. This heat can be harvested and harnessed to generate steam to produce electricity and to provide hot water and heating for buildings.

1.10   Summary

In order to reduce our dependence on exhaustible fossil fuels and reduce excessive greenhouse emissions, we must switch to renewable energy resources, such as wind, solar, and marine. In general, energy can be categorized into four basic types: kinetic, gravitational, electrical, and nuclear. Energy cannot be created or destroyed but can be converted from one form into another. Whenever energy is converted from a low-quality form into a high-quality form, there is a maximum physical limit on the efficiency of the conversion process. Apart from tidal and geothermal, renewable energy resources are solar in principle.

1.11   Problems

P1.1   An object of mass 800 kg is moving in a straight line with a steady speed of 25 m/s. Determine its kinetic energy.

P1.2   An object of mass 5 kg is placed on a table. A force of 200N is applied to the object at an angle of 60° from the horizontal and moves the object 0.5m in the horizontal direction. Calculate the work done in moving the object.

P1.3   Jack takes his 10-kg suitcase up four flights of stairs for a total vertical distance of 20 meters. How much work was done?

P1.4   An object of mass 500 kg is lifted 10m above ground. What is its potential energy?

P1.5   How much electrical energy is required to run a 3-kW oven for 2 hours? Express your answers in kWh and in joules. If the cost of 1 kWh is 15 cents, what is the total cost of this energy?

P1.6   If each unit (kWh) of electrical energy costs 12 cents, how much does it cost to run a 300W TV for 45 minutes?

P1.7   Explain briefly what is meant by renewable energy source and renewable energy technology. Give some examples.

P1.8   A 5-MW wind turbine produces on average 15,000 MWh of electricity in a year.

a. What is the capacity factor of the turbine?

b. On average, for how many hours in a year the turbine runs at full capacity?

P1.9   A 5-MW wind turbine runs at full capacity for a total of approximately 2,500 hours per year.

a. What is the annual amount of electricity generated?

b. What is the capacity factor of the turbine?

P1.10 Explain what is meant by the following terms: primary energy, delivered energy, and useful energy.

P1.11 The capital cost of a proposed 2-MW wind turbine plant is estimated at $1800 per kW of its rated capacity and it is estimated that on average, it will be running at full capacity for 25,00 hours per year. What is the capacity factor of the plant? Making simple calculations, determine the number of years the plant will recover the capital cost assuming the cost of a competing fossil fuel plant is 12 cents per kWh.

P1.12   The cost of one 10W light emitting diode (LED) lightbulb is $10 and has an average life span of 25,000 hours. A 60W incandescent lightbulb provides the same amount of light and costs only $1 and has an average life span of 12,000 hours. Calculate the total operational cost for each bulb over 20 years assuming each bulb is used for an average of 10 hours per day and that the cost of one kWh is 10 cents.

P1.13   What are the advantages and disadvantages of nuclear power generation?

P1.14   Is nuclear energy renewable energy or not? Explain why.

P1.15   Explain the adverse effects of fossil fuels on the environment.

P1.16   Consider the statement "most renewable energy resources are solar in principle." Which resources are not solar and why?

References

[1]   Twidell, J., and A. Weir, *Renewable Energy Resources*, London: Taylor and Francis Group, 1986.

[2]   World Bank, "Fossil Fuel Energy Consumption," 2014, https://data.worldbank.org/indicator/EG.USE.COMM.FO.ZS.

[3]   The Open University, *Renewable Energy Power for a Sustainable Future*, G. Boyle, Ed., Oxford: Oxford University Press, 2012.

[4]   Burshstien, A. I., *Introduction to Thermodynamics and Kinetic Theory of Matter*, New Jersey: Wiley-VCH, 2005.

[5]   US Energy Information Administration, "Annual Energy Review," US EIA, Washington DC, 2011.

[6]   Windpower, "The Betz Limit," http://www.wind-power-program.com/betz.htm.

[7]   Polman A., and H. A. Atwater, "Photonic Design Principles for Ultrahigh-Efficiency Photovoltaics," *Nature Materials*, Vol. 11, 2012, pp. 174–177.

[8]   BP, "BP Statistical Review of World Energy 2018," June 2017, [Online], Available: https://www.bp.com/content/dam/bp/business-sites/en/global/corporate/pdfs/energy-economics/statistical-review/bp-stats-review-2018-full-report.pdf. [Accessed 21 April 2019].

[9]   IEA, "BP Statistical Review of World Energy," BP, June 2018, https://www.bp.com/content/dam/bp/business-sites/en/global/corporate/pdfs/energy-economics/statistical-review/bp-stats-review-2018-full-report.pdf.

[10]   BP, "Statistical Review of World Energy," BP, 2010, http://www.bp.com.

[11]   IEA, "World Energy Outlook 2010," IEA, Paris, 2010.

[12]   Department of Economic and Social Affairs, "Energy Statistics," United Nations, 2018, https://unstats.un.org/unsd/energy/pocket/2018/2018pb-web.pdf.

[13]   International Energy Agency, "Solar Energy," 2018, https://www.iea.org/topics/renewables/solar/.

[14]   World Energy Council, "Hydropower," World Energy Council, 2018, https://www.worldenergy.org/data/resources/resource/hydropower/.

[15]   Renewable Energy Policy Network for the 21st Century, "Renewables 2018," http://www.ren21.net/wp-content/uploads/2018/06/17-8652_GSR2018_FullReport_web_-1.pdf.

[16]   National Geographic, "Five Striking Concepts for Harnessing the Sea's Power," https://news.nationalgeographic.com/news/energy/2014/02/140220-five-striking-wave-and-tidal-energy-concepts/.

[17]   UK Energy Research Centre, "Energy from Biomass: The Size of the Global Resource," Imperial College, London, 2011.

[18]   National Oceanic and Atmospheric Administration, "Tides and Water Levels," U.S. Department of Commerce, https://oceanservice.noaa.gov/education/tutorial_tides/tides02_cause.html.

[19]   National Geographic, "Geothermal Energy," https://www.nationalgeographic.com/environment/global-warming/geothermal-energy/.

2

Photovoltaics: Background Material

The aim of this chapter is to review the basic theory of semiconductor physics and the operation of the pn-junction diode, which are important for understanding the operation and modelling of photovoltaic systems.

2.1 Learning Outcomes

After actively engaging with the material in this chapter, you should be able to

1. Distinguish between intrinsic and extrinsic semiconductor materials;
2. Explain the concept of potential barrier in a static electric field;
3. Describe the operation of a pn-junction diode under forward- and reverse-bias conditions;
4. Solve diode circuit problems using the load-line method;
5. Solve diode circuit problems using the trial and error method;
6. Solve diode circuit problems using large signal equivalent circuits of the diode.

2.2 Overview

This chapter provides the necessary background material covering the basic physics of the semiconductor diode that is relevant to understanding the operation and modeling of PV systems. The chapter explains how the process of doping is used to produce n-type and p-type semiconductors while simultaneously increasing their electrical conductivity. It moves on to explain how p-type and n-type semiconductors are used to form a pn-junction, which is fundamental to the operation of all semiconductor devices, such as PV cells and transistors. The structure and operation of the pn-junction, commonly known as a diode, is explained. The solution of simple diode circuits using graphical, numerical, and approximate large signal circuit models of the diode are presented with numerical examples. Finally, a number of problems are included at the end of the chapter to enhance and consolidate understanding of the material.

2.3   Introduction

In a solar electric power generation system, known as a PV system, the energy in the sunlight is converted to electrical energy making use of the PV effect. The PV effect refers to the phenomenon of generating electricity from light, more precisely from the energy of electromagnetic photons from the sun. Hence, any device or material that can convert light energy into electrical energy is qualified as PV. The most basic photovoltaic device is the PV cell, which is typically fabricated using a semiconductor material such as silicon (Si). The PV cell is a device that is similar in structure to the pn-junction diode. However, a single PV cell provides a limited terminal voltage, typically about 0.5V, and therefore several cells are normally manufactured in a single integrated module, known as a PV panel, in which all cells are connected in series to provide higher terminal voltage. Further, to provide larger amount of power, panels are connected in series and or parallel combinations making up a PV array. The terminal I-V (current-voltage) and power-voltage (P-V) characteristics of a PV array vary dynamically with variations in the ambient temperature and irradiance (i.e., the incident light energy per square meter) [1]. Moreover, these characteristics can also change substantially with physical obstructions, such as clouds, trees, and buildings, which prevents individual panels in an array and or individual cells in a panel, receiving the same amount of sunlight. This phenomenon is known as partial shading [2]. The combined effects of partial shading and variations in irradiance and temperature can drastically reduce the energy yield of a PV system [3]. In order to reduce the impact of these effects, specialized electronic systems have to be designed to optimize the performance of a PV array or panel under varying conditions of temperature, irradiance, and partial shading. However, this requires modeling of the PV cell, panel, and array using lumped circuit parameters. Since the physical structure and characteristics of a PV cell are similar to those of the semiconductor diode, it is not surprising to find that all equivalent circuit models of a PV cell reported in the literature involve one or two diodes [4]. Further, these models, with slight modifications, are also used to model complete PV panels and arrays [5]. Therefore, it is imperative that the designer of a PV system is conversant with the basic physics that underlies the operation of the semiconductor diode. This requires some basic understanding of electrostatic fields and forces that we will consider next. This material is provided for students, scientists, and engineers who would like to learn about PV systems but don't have the necessary background provided in this chapter.

2.4 Coulomb's Law

Coulomb's law states that the force F, measured in newtons (N) between two-point charges Q_1 and Q_2 and measured in coulombs (C), is proportional to the product of the two charges and is inversely proportional to the square of the distance r, in meters (m), between them. That is, the magnitude of this force is

$$F = k\frac{Q_1 Q_2}{r^2}$$

(2.1)

Where k is the constant of proportionality that depends on the medium in which the two charges exist and in free space or vacuum; its value is 9.0×10^9 N.m²/C². The direction of this force is along the line joining the two charges as indicated in Figure 2.1

It is directed outward (i.e., repulsive force) if the two charges are of the same sign, or inward (i.e., attractive force) if the two charges are of opposite signs.

Example 2.1

Two equal point charges of 1 μC each of opposite polarities are fixed rigidly in air 1 mm apart. Calculate the magnitude of the electrostatic force between them.

Solution:

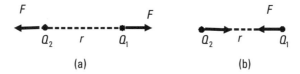

(a)   (b)

Figure 2.1 Two-point charges of the same sign repel (a) and of the opposite sign attract (b).

$$F = k\frac{Q_1 Q_2}{r^2}$$

$$F = 9.0 \times 10^9 \frac{(1 \times 10^{-6})(1 \times 10^{-6})}{(1 \times 10^{-3})^2} = 9000 \text{ N}$$

This is a large force sufficient to lift a small car.

2.5 Static Electric Field

Consider a positive point charge Q (C) placed firmly at a point in air as shown in Figure 2.2. If another positive point charge q is brought into the vicinity of Q, it will be acted upon by a force. This coulomb's force is directed radially outward (i.e., it is repulsive) and its magnitude increases as q is brought closer to Q as expressed by the varying lengths of the arrows. It may be said that Q has an electric field around it where electric forces act on charged objects. Since the source of the field is a stationary electric charge (i.e., Q), such a field is referred to as a static electric or electrostatic field. Quantitatively, an electric field is expressed in terms of the electric field intensity, E, which is defined as the force per unit charge. That is

$$E = \frac{F}{q}$$

$$(2.2)$$

Substituting for the force F from (2.1), we can write:

$$E = k\frac{Q}{r^2} \tag{2.3}$$

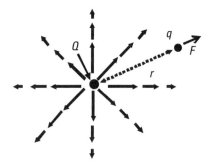

Figure 2.2 Point charge Q with vectors indicating magnitude and directions of the associated electric field.

The electric field intensity, E, also known as the electric field gradient or strength is a vector quantity whose magnitude is given by (2.3) and at any point in the field, its direction is the direction of the force experienced by a small positive test charge placed at that point. The test charge must be small enough so that it does not disturb the field being measured. Since the smallest physical charge is the charge of an electron, an electric field cannot be measured with absolute certainty.

2.6 Static Electric Potential

The electric field of the point charge, described above is a nonuniform field. A uniform static electric field may be obtained between two parallel plates charged with opposite polarities as shown in Figure 2.3(a). Now consider two points, points x_1 and x_2, in a uniform electric field E whose lines are parallel to, and in the direction of, the positive x-axis and have the same magnitude everywhere in the region of interest as shown in Figure 2.3(b). Let a positive test charge at x_2 be moved in the negative x-direction to (i.e., opposite to the direction of the field). Clearly, the field exerts a force on the charge in the positive x-direction therefore, it requires work (i.e., energy) to move the charge against the force exerted by the field.

Work done is defined as

$$\text{Work done} = \text{Force} \times \text{Distance}$$

and

$$\text{Work done per unit charge} = \text{Force per unit charge} \times \text{Distance}$$

Dimensionally, this is

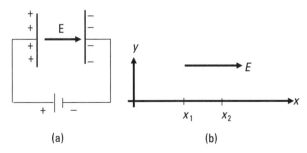

(a)   (b)

Figure 2.3 (a) Uniform electric field, and (b) linear path of charge in a uniform electric field.

Joules per coulomb = (Newtons/Coulomb) × (meter)

However, the force per unit charge is simply the electric field intensity E. Hence, the work done in moving a unit of charge between the two points x_1 and x_2 is

$$W = E\left(x_2 - x_1 \right) \left(J/C \right)$$

(2.4)

We define the static potential difference V, between two points in a static electric field as the work done in transporting a unit of charge between the two points. That is

$$V = E(x_2 - x_1)$$

(2.5)

Point x_1 is said to be at a higher potential than point x_2 since it requires work to reach it from x_2. Thus, moving opposite to the field, we experience a rise in potential. The unit of electric potential, V, is the volt (V) and 1V is equal to 1 joule per coulomb. Consequently, in a uniform electric field, the potential difference between any two points, d (m) apart is given by

$$V = Ed$$

(2.6)

Note that the electric field intensity E is expressible in either newton per coulomb or in volt per meter.

Since the potential difference, V, between two points in an electric field, is the amount of energy required to move a unit of charge, then to move q coulombs of charge, the work done in joules is

$$W = qV \quad (J)$$

(2.7)

2.7   The Concept of Static Potential Barrier

The concept of the static potential barrier is very important for understating the operation of the pn-junction. Consider a uniform electric field as shown in Figure 2.4. Clearly, an electron on the negative plate will be accelerated by the electric field and move to the positive plate. In this case, we say that the electron gained energy from the field, or work has been done on the electron by the field. Suppose that the potential difference between the two plates is 1V, then the energy gained by the electron is the charge on the electron times the potential difference (i.e., $W = 1.6 \times 10^{-19}$) joules. This is a very small number; hence,

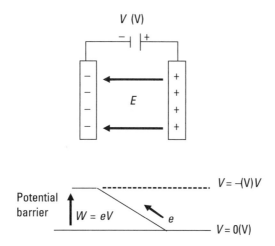

Figure 2.4 The concept of potential barrier.

when dealing with small charges, we use a different unit for measuring energy, called the electron-volt (eV). Since the electron has been moved through a potential difference of one volt, we say that it has gained one eV of energy. That is,

$$1 \text{ eV} = 1.6 \times 10^{-19} \text{ joules} \tag{2.8}$$

Now suppose we want to move the electron from the positive plate to the negative plate (i.e., against the force exerted on it by the electric field). In this case, work has to be done against the field. This is like climbing up a hill; indeed, we say that there is a potential barrier that the electron has to climb. However, to climb this barrier, the electron needs some energy that is at least equal to the energy presented by the barrier (i.e., $W = eV$). Compare this situation with the potential barrier in a pn-junction, which we will be considering soon.

2.8 Intrinsic Semiconductors

Metals, such as copper and aluminium, are good conductors of electricity because they have mobile charge carriers (i.e., free electrons, which are responsible for conduction of electric current). In metals, electrons in the outer shell of an atom, known as valence electrons, are weakly bound to their parent atom. Hence, in a conductor, there might be one or two or three free electrons per atom that can break away from their parent atom and become free electrons (i.e., free to move around the body of the metal). These electrons are said to have left the valence band and moved into the conduction band. Those mobile charge carriers move in totally random fashions; hence even when a closed circuit is provided, they cannot produce an electric current. However, with a closed circuit and an external source of energy, such as a battery, these free electrons can be made to move in one direction constituting electric current. The electrical conductivity of a material is proportional to the concentration n of free electrons. For a good conductor, n is in the order of 10^{28} electrons/m^3 [6]. For a good insulator n is typically less than 10^7 electrons/m^3. For intrinsic (i.e., pure) semiconductors, the concentration of free electrons lies between these two values. In a pure semiconductor, such as silicon, each atom has four valence electrons and each valence electron in each atom is shared by one of the immediate four neighbors as depicted in Figure 2.5. This electron-pair, which is known as a covalent bond, is represented by the dashed lines that join each atom to its neighbor. This covalent bond results in a strong binding force between an electron and the nucleus of the atom.

2.9   Holes and Electrons as Charge Carriers

At low temperatures near zero, pure semiconductors behave like insulators. However, at higher temperatures greater than room temperature, some covalent bonds will be broken by the acquired thermal energy. Once a covalent bond is broken an electron is dislodged from its location, as depicted in Figure 2.6. This electron is now free to wander in the interior of the semiconductor crystal and

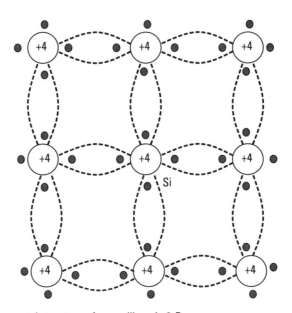

Figure 2.5   The crystal structure of pure silicon in 2-D.

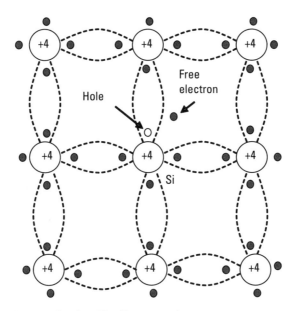

Figure 2.6 A broken covalent bond in silicon crystal.

is available as a free charge carrier (i.e., available for current conduction). As shown, a breakaway electron leaves an uncovered or incomplete bond called a hole. A hole represents a situation where it is easy for an electron from a neighboring atom to move in and occupy it. However, this electron has also left a broken bond or equivalently a hole. Consequently, another electron will move from another atom to fill it, leaving a new hole, and so on. This means that as electrons move in one direction, holes move in the opposite direction.

As far as electric current is concerned, holes behave like positive charge carriers and this means that electrons and holes moving in opposite directions contribute to the same direction of current. In a pure semiconductor, the number of holes is the same as the number of free electrons. Every time thermal energy produces an electron-hole pair, another electron-hole pair is lost due to recombination of a free electron and a hole. Hence, the semiconductor remains electrically neutral. A free electron in a semiconductor is said to have acquired sufficient energy to move from the valence band (where it cannot conduct electric current) to the conduction band where it becomes free and available to conduct electric charge. When an electron moves from the valence band to the conduction band, it requires an amount of energy equal to the bandgap energy between the two bands, which depends on the semiconductor material. The bandgap energy is measured in eV, and for silicon, it is approximately 1.12 eV. This means that an electron needs to gain 1.12 eV so that it can leave its own parent nucleus and move to the conduction band as a free electron able to contribute to current flow. In a photovoltaic cell, this energy comes to the electron from the energy in the sunlight. In a normal diode, this energy comes from an external voltage source (e.g., a battery).

2.10   Extrinsic Semiconductors

The conductivity of intrinsic semiconductor can be improved by the process of doping, which involves adding impurities to the intrinsic semiconductor. The new semiconductor becomes extrinsic (i.e., impure) semiconductor. Extrinsic semiconductor can be one of two types: n-type or p-type semiconductor depending on the type of the added impurities (i.e., the dopant). If the dopant is a pentavalent (i.e., has five valence electrons, such as phosphorus (P) and arsenic (As)) then four of the five valence electrons will form four covalent bonds with the host semiconductor atoms. This leaves one electron free as a charge carrier, as depicted in Figure 2.7. The impure (i.e., doped) semiconductor is known as n-type semiconductor because the majority charge carriers in it are electrons. It will have a few holes due to thermal agitation and we refer to these as the minority charge carriers. The pentavalent atoms added to the semiconductor are referred to as donor atoms because they donate electrons to the host silicon.

If on the other hand, pure semiconductor is doped with a trivalent material (i.e., one with three valence electrons), such as indium (In) or boron (B), as shown in Figure 2.8, then only three of the covalent bonds can be filled. This leaves a vacancy in the fourth bond , which is equivalent to a hole (i.e., a positive charge carrier). This hole can accept an electron from the host semiconductor;

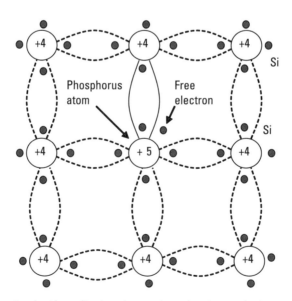

Figure 2.7   A pentavalent impurity phosphorus atoms donates an electron.

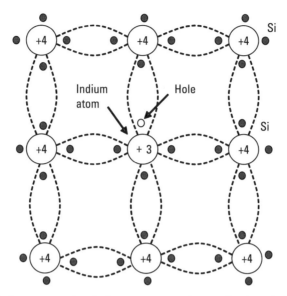

Figure 2.8 A trivalent atom such as indium accepts an electron, leaving a hole.

hence, the dopant atoms are called acceptors. The doped semiconductor is referred to as a p-type semiconductor because the majority charge carriers are holes and the thermally generated minority charge carriers are electrons.

When pure semiconductor is doped with an impurity of one part (dopant) to 10^8 (semiconductor), the conductivity of the doped semiconductor near room temperature is multiplied by a factor of 24,100 [6]. Thus the process of doping not only results in the useful p-type and n-type semiconductors, but also increases the conductivity of pure semiconductor by considerable margin.

2.11   Pn-Junction Diode

A p-type (or n-type) semiconductor on its own behaves like a normal conductor that can conduct current in either direction. However, when n-type and p-type semiconductors are joined together, as shown in Figure 2.9, they form a junction known as a pn-junction diode, or simply just a diode, whose electrical characteristics are central to the operation of all semiconductor devices.

The most important characteristic of a diode is its ability to allow current flow only in one direction (i.e., it acts as a rectifier). When donor atoms are introduced into one side and acceptor atoms into the other side of a semiconductor crystal as depicted in Figure 2.9, a pn-junction is formed. A donor atom is represented by a plus sign because after a donor atom donates an electron it becomes a positive ion. Similarly, an acceptor atom is represented by a minus sign because it becomes a negative ion after it accepts an electron. Initially, (i.e.,

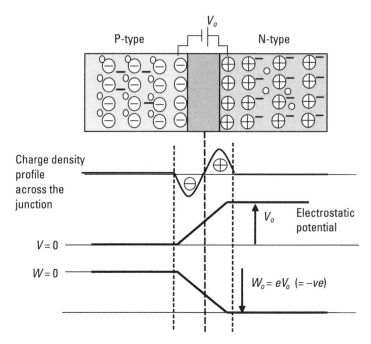

Figure 2.9 From top to bottom: A pn-junction diode, the charge density distribution, the electrostatic potential, and the potential barrier W for electrons.

at the instant the junction is formed) there are mainly electrons in the n-type region and mainly holes in the p-type region.

However, because there is a charge density gradient across the junction, holes from the p-type region (where they are the majority charge carriers) will start to diffuse to the right across the junction (i.e., to n-region where they are minority carriers), see Figure 2.9. At the same time electrons from the n-type region (where they are the majority charge carriers) start to diffuse to the left (i.e., to the p-type region where they are minority carriers). This diffusion process cannot continue forever. Why? Well, in the n-type region positive donor ions are neutralized by the majority electrons, and in the p-type region, negative acceptor ions are neutralized by the majority holes; hence, overall the crystal is neutral. However, once some free electrons near the junction leave the junction and cross into the p-type region, they leave behind unneutralized positive donor ions near the junction. At the same time, positive holes from the p-type region cross the junction leaving behind unneutralized negative acceptor ions. Those migrating electrons and holes combine to form electron-hole pairs, but with the result that an electrostatic potential is developed across the junction. This electrostatic potential creates an electrostatic field across the junction whose direction is from right to left. That is the electrostatic field prevents any further electrons crossing the junction from the n-type to the p-type region and prevents any holes crossing the junction from the p-type region into the n-type region. Therefore, the region around the junction becomes depleted of mobile charge carriers; hence, it is called the depletion layer. The width of this layer is typically 0.5 μm in silicon [6]. In other words, majority carriers cannot cross the junction without energy from an external source. For an electron to cross the junction from the n-region into the p-region, an electron must overcome the potential barrier created by this electrostatic potential. The same applies to holes trying to cross the junction into the n-type region. This energy must come from an external source, such as a battery.

2.11.1 Pn-Junction under Reverse-Bias Condition

If an external voltage source such as a battery, V_B, is connected to a pn-junction so that the negative terminal is connected to the p-region, which is called the anode, and the positive terminal is connected to the n-region, which is called the cathode, as shown in Figure 2.10, the pn-junction is said to be reverse-biased. The polarity of the external source is such that it is in the same direction as the electrostatic potential V_o and therefore, it causes holes and electrons to move away from the junction, increasing the width of the depletion layer and the height of the potential barrier. Therefore, the flow of majority carriers is further reduced and nominally no forward current is possible, since forward current is due to the flow of majority carriers; that is, electrons from the n-type region crossing the junction to the p-type region and holes from the p-type region crossing the junction to the n-type region. However, due to thermal agitation, some electrons-hole pairs are generated throughout the structure. In the n-type these thermally generated holes are referred to as the minority carriers, and similarly, the minority carriers in the p-region are the electrons.

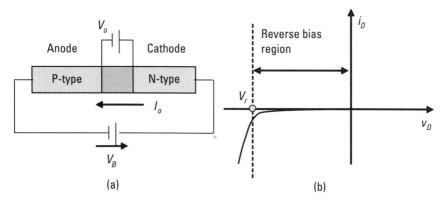

Figure 2.10 (a) Pn-junction under reverse-bias and physical structure, and (b) i-v characteristic.

Holes in the n-region near the junction will be pulled across the junction by the electrostatic field into the p-region. Similarly, the minority electrons in the p-region will be pulled across the junction into the n-region. Note that holes and electrons traveling in opposite directions give rise to the same current. This current that is due to the minority charge carriers crossing the junction is called the reverse saturation or leakage current I_O and is very small, typically in the range of 10^{-10} to 10^{-12} A. Note that by convention, the positive direction of current flow is the direction of positive charge carries (i.e., holes). This is opposite from the direction of real current, which is the direction of negative charge carriers (i.e., electrons). Hence, the conventional direction of the reverse saturation current is from the cathode to anode. The reverse saturation current of a diode is an important diode parameter, which is specified in the manufacturer's datasheet of the diode. In the reverse-bias, also known as the reverse blocking mode of operation, the diode exhibits very large resistance (in the order of millions of ohms). The terminal i-v characteristic indicates that as the reverse voltage is increased, the reverse saturation current remains almost constant. Indeed, the reverse saturation current only depends on temperature and is not dependent on the reverse-bias voltage. However, if the reverse-bias voltage is increased beyond a certain value that is a characteristic of a given diode known as the reverse breakdown voltage V_r, the diode enters the breakdown region and will be permanently damaged. The breakdown voltage is typically greater than 50V for silicon diodes and can be found in the manufacturer's datasheet of the diode. However, some diodes are deliberately manufactured to operate in the breakdown region, for example the zener diode, which is used as a voltage regulator. More about these and other types of didoes can be found in electronics books, such as [6, 7].

2.11.2   Pn-Junction under Forward-Bias Condition

If an external source, such as a battery, V_B, is connected to a pn-junction so the negative terminal is connected to the cathode (the n-region) and the positive terminal is connected to the anode (the p-region) as shown in Figure 2.11, the pn-junction is said to be forward-biased. Under forward-bias conditions, the external voltage opposes the electrostatic potential and; hence, lowers the potential barrier. The depletion layer is narrowed and holes from the p-type cross the junction into the n-type where they constitute injected minority charge carriers. Similarly, electrons from the n-type cross the junction into the p-type where they constitute injected minority charge carriers. Holes traveling from left to right result in a current in the same direction as electrons traveling from right to left. That is, the resultant current crossing the junction is the sum of the holes and electrons currents. This is the forward component of the diode current and by convention it is in the direction of the holes' flow (i.e., from the

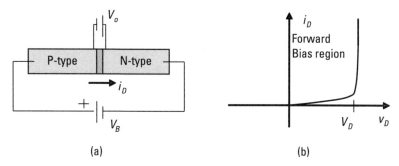

Figure 2.11 (a) Pn-junction under forward-bias condition, and (b) i-v characteristic.

anode to cathode). In the forward-bias mode, the diode exhibits small resistance ranging from few to about 25 ohms depending on the current.

The forward i-v characteristic indicates that a small threshold voltage appears across the diode before it starts to conduct. This voltage V_D is approximately constant for a given diode, although it varies slightly with the diode current, and is in the range of 0.5V to about 0.8V for silicon diodes, and typically a value of 0.7V is used in the analysis and design of diode circuits. The variation of the forward diode current with the voltage across the diode is given by the exponential diode model that closely approximates the real characteristic of a diode:

$$i_{DF} = I_o e^{\frac{v_D}{nV_T}} \tag{2.9}$$

In this equation, I_o is the reverse saturation current, n is a number between 1 and 2 for silicon diodes called the ideality factor, e is the base of the natural logarithm, and v_D is the voltage drop across the diode. The voltage V_T is called the thermal equivalent voltage, which is given in terms of the Boltzmann's constant $k = 1.381 \times 10^{-23}$ (m²kg s^{-2}K^{-1}), room temperature T in kelvin (K), and the electronic charge q = 1.6×10^{-19} (C) as

$$V_T = \frac{kT}{q} \tag{2.10}$$

At room temperature of 293 kelvin, this is about 25 mV. The other current component is the reverse saturation current, which depends on temperature. Hence, the net diode current is the difference between the forward current and the reverse saturation current and this is given by the exponential model

$$i_D = I_o e^{\frac{v_D}{nV_T}} - I_o = I_o (e^{\frac{v_D}{nV_T}} - 1) \tag{2.11}$$

In the forward-bias mode, however, the exponential term becomes much larger than unity; hence, we tend to model the net diode using the approximate equation

$$i_D = I_o e^{\frac{v_D}{nV_T}}$$

(2.12)

Since the pn-junction diode possesses very high resistance when reverse-biased and small resistance when forward-biased, it has the property of a rectifier and indeed it is also known as a rectifier. Although we have shown that a battery is connected directly across the diode when forward-biased, this was only for conceptual purposes. In practice a voltage source should never be connected directly across a diode; otherwise, a large current will flow causing excessive heat that will damage the diode. A resistance must always be connected in series with a diode to keep the current below its maximum rated value specified in its datasheet.

2.12   Semiconductor Diode as a Circuit Element

The circuit symbol of a diode and a typical terminal i-v characteristic are shown in Figure 2.12. In addition to the general use of diodes as a rectifier, diodes have a multitude of circuit applications, such as signal limiters, dc level shifters, signal mixers, demodulators, and filters. There are many types of diodes, for example the LED, which when forward-biased emits light, and photodiodes that convert light into an electrical signal and are used as optical demodulators and laser diodes.

Laser diodes are used to convert an electrical signal into a laser (i.e., coherent light, and they are used in optical communications as modulators). Before we can analyze or design electronic circuits that involve diodes, we need to learn some techniques of solving diode circuits, which is the subject of the next section.

2.13   Solving Diode Circuit Problems

Clearly, a diode is a nonlinear circuit element (i.e., it does not obey Ohm's law) and we need to learn special techniques for analyzing electric circuits with diodes. These techniques use either iterative methods or involve some approximations, which make the analysis easier; in addition, we will study a simple graphical method. Each method will be illustrated with an example. Consider the circuit shown in Figure 2.13 (a) and let us try to determine the circuit current. We can see that the source voltage E is the sum of the voltage across the resister $V_R = I_D R$ and the voltage across the diode V_D. This is a statement of

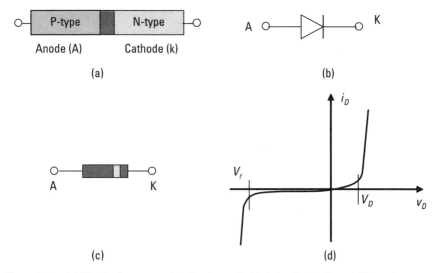

Figure 2.12 (a) Physical structure, (b) circuit symbol, (c) physical look, and (d) i-v characteristic of a diode.

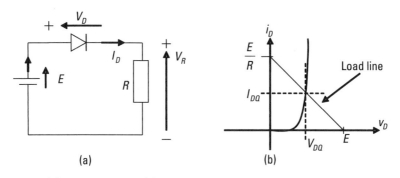

Figure 2.13 (a) Diode circuit and (b) load-line.

Kirchhoff's voltage law (KVL), which states that around a closed loop, the sum of the voltage rises (i.e., sources of energy, such as batteries) must equal the sum of the voltage drops (i.e., sinks of energy, such as IR); that is

$$E = I_D R + V_D \qquad (2.13)$$

This equation looks simple enough until we realize that it involves two unknowns: the diode current I_D and the diode voltage V_D. Hence, we cannot solve it unless, of course, we want to use an approximate value of the diode voltage (e.g., 0.7V). Although this can be acceptable in many circuit applications, there are others that require an accurate value of this voltage. Hence, to find the diode voltage and current we clearly need a second equation. This second equation is the diode exponential model (i.e., the equation)

$$i_D = (I_O e^{\frac{v_D}{nV_T}} - 1) \qquad (2.14)$$

There are many ways of using the above two equations to find the diode voltage and current. However, we will discuss only two methods here: a graphical method, called the load-line method, and an iterative technique, known as the trial and error method.

2.13.1   Load-Line Method

To solve the above problem using the load-line method, we need the terminal (i-v) characteristic of the diode, which is normally obtained by experiment or from the manufacturer's datasheet. This is basically a plot of the diode i_D (dependent variable) against the diode voltage v_D (independent variable), which is closely modeled by the diode exponential equation, Figure 2.13(b). We write the voltage loop equation

$$E = i_D R + v_D$$

Rearranging, to make the current the dependent variable, and dividing by R

$$i_D = \frac{E}{R} - \frac{1}{R} v_D$$

This is now an equation of a straight line, called the load-line, which we can plot on the (i-v) axes, as shown in Figure 2.13 (b). Setting the voltage $V_D = 0$, we can determine the vertical intercept of the load-line as ($i_D = E/R$). Similarly, when the current is zero $i_D = 0$, we can find the horizontal intercept as $v_D = E$. The gradient of the load-line is $-1/R$. The solution point, also called the quiescent or the operating point, (V_{DQ}, I_{DQ}) is the point of intersection of the load-line and the (i-v) characteristic.

Example 2.2

Use the load-line method to determine the current through, and the voltage across, the diode D in the circuit of Figure 2.14, whose (i-v) characteristic is also included in the figure.

     Solution:

     The load-line equation is

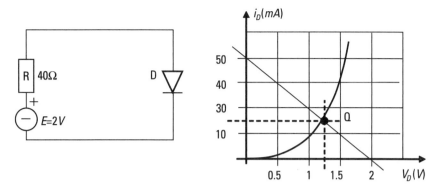

Figure 2.14 Circuit diagram and i-v characteristic for Example 2.2.

$$i_D = \frac{E}{R} - \frac{v_D}{R}$$

Substituting values, we obtain

$$i_D = \frac{2}{40} - \frac{1}{40} v_D$$

To find the current (vertical axis) intercept, we substitute $v_D = 0$ in the load-line equation; hence, the load-line intersects the current axis at the point $I_D = \frac{2}{40} = 50 \, mA$.

To find the intersection of the load-line with the voltage axis, we substitute $i_D = 0$ in the equation of the load-line:

$$0 = \frac{2}{40} - \frac{v_D}{40}$$

Hence, the load-line intersects the voltage axis at the point $v_D = 2\,V$. Now we draw the load-line and we find the point of intersection with the diode characteristic as

$$(V_{DQ}, I_{DQ}) = (1.25 \, V, 15 \, mA)$$

This is the solution, or the quiescent point.

2.13.2 Trial and Error Method

To illustrate this method, we will assume that $n = 1$ and the diode is at room temperature. In fact, we will always use this assumption, unless told otherwise. In addition, since the reverse saturation current is negligible compared to the main diode current, we will use the simplified diode equation

$$i_D = I_0 e^{\frac{v_D}{0.025}} = I_0 e^{40 v_D} \qquad (2.15)$$

We start the solution procedure by assuming a value for the diode voltage. This is called the first guess, say V_{D1}. We substitute this voltage into the above current equation and determine the diode current, say I_{D1}. We then substitute these two values (i.e., the diode voltage V_{D1} and current I_{D1} in the voltage equation $E = I_D R + V_D$). If the equation holds; that is, if the sum $I_D R + V_D$ equates to the supply voltage, E within a small and acceptable error, then our first guess of the diode voltage is correct; otherwise we have to adjust our initial guess, either increase or decrease it, and start again. We repeat this iterative process until the error in the voltage is acceptable (i.e., the difference between $I_D R + V_D$ and E is small enough). This method is best illustrated with an example.

Example 2.3 Trial and Error Method

A diode is connected in series with a resistor R of 10 ohms across a 5V DC source, so that the diode is forward-biased. Given that $V_T = 25 mV$, $n = 1$, $I_O = 10^{-12} A$. Use the trial and error method to determine the diode current and voltage.

Solution:

We write the loop equation using KVL as

$$E = I_D R + V_D \qquad (A)$$

and we have the exponential model of the diode

$$I_D = 10^{-12} e^{40 V_D} \qquad (B)$$

Let us start with a guess of, $V_D = 0.6 V$ and substitute this value in Equation B

$$I_D = 10^{-12} e^{40 (0.6)} = 0.0265 \text{ A}$$

Now substitute these values of diode voltage and current Equation A

$$I_D R + V_D = 0.0265(10) + 0.6 = 0.8650 \text{ V}$$

This is too small compared to the value of E of 5V. Hence, we increase our guess of the diode voltage, say $V_D = 0.65\,V$ and repeat the above process:

$$I_D = 10^{-12}\,e^{40\,(0.65)} = 0.1957\;A$$

Again, substitute these values of diode voltage and current Equation A

$$I_D R + V_D = 0.19573(10) + 0.65 = 2.610\;\;V$$

This is still smaller than the required 5V, so let us try $V_D = 0.7\;V$

$$I_D = 10^{-12}\,e^{40\,(0.7)} = 1.4463\;A$$

Substitute these values of diode voltage and current Equation A:

$$I_D R + V_D = 1.44623(10) + 0.7 = 15.163\;V$$

This is much larger than the required 5V. Hence, we try a smaller value, something between the two previous values, say $V_D = 0.675\,V$

$$I_D = 10^{-12}\,e^{40\,(0.675)} = 0.5320\;A$$

Hence,

$$I_D R + V_D = 0.5320(10) + 0.675 = 5.9955\;V$$

The error in the voltage is 0.995V (i.e., almost 20%, which is not acceptable). Therefore, we try $V_D = 0.67\,V.$

$$I_D = 10^{-12}\,e^{40\,(0.67)} = 0.4356\;A$$

Hence,

$$I_D R + V_D = 0.4356(10) + 0.67 = 5.026\;V$$

The error in the voltage is now only 26 mV or 0.52%, which is acceptable. Therefore, the DC the quiescent point is (0.67 V, 0.4356 A). If more accurate values are required, then we need to continue the above process of iteration.

Sometimes, we need to determine the voltage drop across a diode given its current. In this case, we can rewrite the above diode equation in terms of the diode voltage

$$I_D = I_O e^{\frac{V_D}{nV_T}}$$

Taking the natural log of both sides

$$\ln(I_D) = \ln(I_O) + \frac{V_D}{nV_T}$$

Rearranging for the diode voltage

$$V_D = \left(\ln(I_D) - \ln(I_O)\right)nV_T \tag{2.16}$$

Example 2.4

A forward-biased silicon diode has a reverse saturation current of 10 pA and a current of 100 mA. Assuming that its ideality factor is 1 and the thermal equivalent voltage is 25 mV, determine the forward voltage drop across the diode.

$$\begin{aligned} V_D &= \left(\ln(I_D) - \ln(I_O)\right)nV_T \\ &= \left(\ln(0.1) - \ln(10^{-12})\right)(1 \times 0.025) \\ &= 0.6332 \ V \end{aligned}$$

2.14   Large Signal Diode Models

In many practical applications, the sort of accuracy obtained using the above graphical and numerical methods is not required. In such cases, we use simplified equivalent circuit models of the diode known collectively as the large signal or DC models.

2.14.1   The Ideal Diode Model

The ideal diode model assumes that when a diode is forward-biased, it behaves like a closed switch with no forward voltage drop and no forward resistance as indicated in Figure 2.15.

    This means that an ideal diode can conduct its full rated current without any voltage drop across it. In the reverse-bias mode, the ideal diode model assumes that the diode behaves like an open switch (i.e., has an infinite reverse resistance) and therefore zero reverse current. This model works fine when a diode is part of a circuit where the external resistance is much larger than the

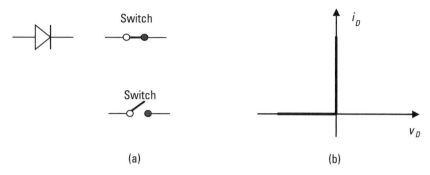

Figure 2.15 Ideal diode model and terminal characteristic.

diode's forward resistance and the external voltages involved are much larger than the diode forward voltage drop, which is typically about 0.7V.

2.14.2 The Battery Plus Ideal Diode Model

When the forward voltage drop across a conducting diode cannot be ignored, we can use the battery plus ideal diode model shown in Figure 2.16. When the diode is conducting, we model it with a battery that emulates its voltage drop, typically 0.7V, in series with the ideal diode model (i.e., a closed switch). This model is used when the external voltage in the circuit is comparable to the diode voltage drop, typically less than 5V, but the external resistance is much larger than the forward resistance of the diode, which is typically less than 25 ohms.

2.14.3 The Battery Plus Resistance Plus Ideal Diode Model

In this model, shown in Figure 2.17, a resistance R_D is included to approximate the forward resistance of the diode.

This is obtained by approximating the forward i-v characteristic into a straight line and calculating the reciprocal of its gradient. This model is

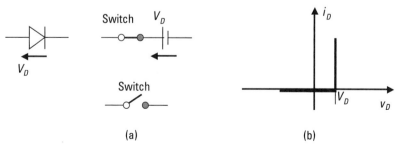

Figure 2.16 Battery plus ideal diode model and terminal characteristic.

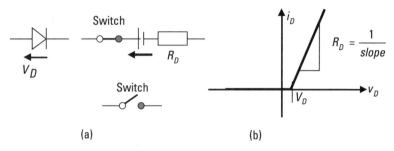

Figure 2.17 Resistance plus battery plus ideal diode model and characteristic.

used when the external resistance and voltage are comparable to the forward resistance and voltage drop of the diode, respectively.

Example 2.5

A DC voltage source of E (V) is connected in series with a silicon diode and a resistor of resistance R ohms. Use an appropriate large signal model of the diode to estimate the circuit current in each of the cases (a)-(c) given below. Note that a negative voltage source here means that the diode is reverse-biased. Also note that the appropriate model does not imply the most accurate.

$$(a) \ \ R = 200 \ \Omega, \ V_{DD} = 8 \ V$$
$$(b) \ \ R = 10 \ \Omega, \ V_{DD} = -10 \ V$$
$$(c) \ \ R = 500 \ \Omega, \ V_{DD} = 5 \ V$$

Solution:

(a) Since both the external resistance and the voltage are much larger than the forward resistance and voltage drop of the diode, we can estimate the current using the ideal diode model. In this case the current is simply 8/200 = 40 mA.

(b) The diode is reverse-biased and we can safely assume that it will not enter the reverse- bias region, hence, the ideal diode model tell us that the current will be zero.

(c) In this case the voltage is not really much greater the diode forward voltage drop so we use the ideal diode plus battery model, giving a current of (5–0.7)/500 = 8.6 mA.

2.15   Problems

P2.1   a. State Coulomb's law.

b. Sketch the electric field distribution around a negative point charge placed firmly in air.

c. Define the electric field intensity of a static electric charge.

d. Define the electrostatic potential difference.

e. Define the eV as a unit of energy.

f. Explain the difference between n-type and p-type semiconductor materials.

g. Describe how the potential barrier is initially established across a pn-junction.

h. Explain the operation of a pn-junction diode under forward- and reverse-bias conditions and how the resulting currents are affected.

P2.2   Calculate the potential difference between two points 2 cm apart in a uniform static electric field of 20 V/m. What is the energy in joules required to move a point charge of 100 mC between the two points? What is this energy in eV?

P2.3   Two electrons are placed firmly in air 1 mm apart. Determine the electric force between them.

P2.4   Explain what is meant by a potential barrier in an electrostatic field. Determine the energy, in joules, needed to move a charge of 1 coulomb through a potential difference of 5V. Express you answer in eV.

P2.5   Given that 50 eV were needed to move a charge through a potential difference of 100V, determine the charge.

P2.6   A silicon diode at 20°C has a reverse saturation current of 10 pA, a forward voltage drop of 0.62V, and an ideality factor of 1. Determine the diode current.

P2.7   A forward-biased silicon diode at 20°C has a reverse saturation current of 5 pA and a current of 200 mA. Assuming that its ideality factor is 1.2, determine the forward voltage drop across it.

P2.8   At 20°C, a silicon diode has a reverse saturation current of 0.5 nA and a forward voltage drop of 0.7V. Determine the diode current given that its ideality factor is $n = 1.5$.

P2.9   At a temperature of 20°C, a silicon diode has a reverse saturation current of 10 pA and an ideality factor of 1. The diode is connected in series with a DC source of 6V and a resistor R of 25 ohms. Use trial and improvement to determine the diode current and voltage drop. Stop the iteration when the error in the voltage is less than 50 mV.

P2.10   A DC series circuits consists of a silicon diode, a resistor of 100 ohms and voltage source of 5V in such a way that diode is forward-biased. The reverse saturation current of the diode is 10 pA and the thermal equivalent voltage is 25 mV. Assuming that the ideality factor is 1, use an iterative method to find the diode current and voltage. Continue the iterative process until the error in the voltage is less than 50 mV.

P2.11   A diode circuit consists of a DC voltage source (V), a resistor R and a silicon diode. Select an appropriate, large-signal diode model to estimate the circuit current in the following cases. You may assume that the average DC resistance of the diode in the forward-bias region is 10 ohms. Note that a negative DC source implies the diode is reverse-biased.

(a) $R = 500\ \Omega$, $V_{DD} = 10\ V$
(b) $R = 100\ \Omega$, $V_{DD} = -5\ V$
(c) $R = 500\ \Omega$, $V_{DD} = 3\ V$
(d) $R = 500\ \Omega$, $V_{DD} = 2.5\ V$
(e) $R = 20\ \Omega$, $V_{DD} = 2.5\ V$
(f) $R = 250\ \Omega$, $V_{DD} = 0.25\ V$

P2.12   A nonlinear device D is connected in series with a 4V DC voltage source and a 50 ohms resistor as shown in Figure 2.18, which also

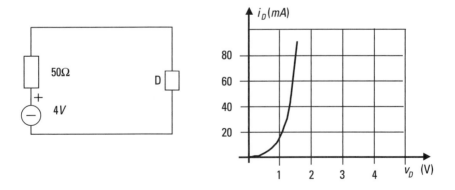

Figure 2.18   The terminal characteristic of the device D.

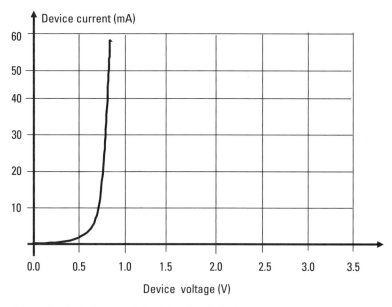

Figure 2.19 Terminal characteristic of the 1N50X diode.

includes the terminal characteristic of the device. Use the load-line method to determine the operating point of the device.

P2.13 The silicon diode 1N50X whose terminal characteristic is given in Figure 2.19 is connected in series with a resistor of 50 ohms. The combination is connected across a DC voltage source of 3V in such a direction to forward bias the diode. Use the load-line method to determine the diode current and voltage.

References

[1] Ibrahim, H., and N. Anani, "Variations of PV Module Parameters with Irradiance and Temperature," *Energy Procedia*, Elsevier , Vol. 134, 2017, pp. 276–85.

[2] Balasubramanian, I. R., S. I. Gansen, and N. Chilakapat, "Impact of Partial Shading on the Output Power of PV Systems under Partial Shading Condtions," *IET Power Electronics*, Vol. 7, No. 3, 201,4 pp. 657–666.

[3] Silvestre, S., and A. Chouder, "Effects of Shadowing on Photovoltaic Module Performance," *Progress in Photovoltaics Research*, Vol. 16, No. 2, 2008, pp. 141–149..

[4] Vinod, N., R. Kumar, and S. K. Singh, "Solar Photovoltaic Modeling and Simulation: As a Renewable Energy Solution," *Energy Reports*, Vol. 4, 2018, pp. 701–712.

[5] Villalva, M. G., J. R. Gazoli, and E. R. Filho, "Comprehensive Approach to Modeling and Simulation of Photovoltaic Arrays," *IEEE Trans Power Electronics*, Vol. 24, no. 5, 2016, pp. 69–78.

[6]　Sedra, A. S., and K. C. Smith, *Microelectronic Circuits*, Northampton: Oxford University Press, 2015.

[7]　Boylestad, R., and L. Nashelsky, *Electronic Devices and Circuit Theory*, Delhi: Pearson, 2013.

3

Photovoltaics: Characteristics and Circuit Modeling of the PV Cell

The aim of this chapter is to introduce the photovoltaic cell as a solar-to-electricity energy converter.

3.1 Learning Outcomes

After actively engaging with the material in this chapter, you should be able to

1. Define the photovoltaic effect;
2. Explain the bandgap energy theory;
3. Use the relationship $E = hf$;
4. Describe the operation of a silicon photovoltaic cell;
5. Define the term air mass (AM) ratio;
6. Define the terms spectral irradiance, insolation, and radiation;
7. Explain the theoretical and practical limitations on the efficiency of semiconductor photovoltaic cells;
8. Analyze a photovoltaic cell using the single-diode equivalent circuit models;
9. Use a spreadsheet to generate and plot the I-V characteristics of a photovoltaic cell.

3.2 Overview

We have learned about semiconductor materials, intrinsic and extrinsic semi-conductors, and considered in some details the operation and modeling of the pn-junction diode. Now we are in a position to study the operation, character-istics, and modeling of the PV cell, which is the fundamental building block of any PV system. The chapter defines the photovoltaic effect and explains the bandgap theory of semiconductors, which is central to the operation of a PV cell. Lumped circuit parameter models of a PV cell based on the single-diode equivalent circuit are also explained. Finally, a number of problems are included at the end of the chapter to enhance and consolidate understanding of the material.

3.3 Introduction

Solar energy, the radiant light and heat from the sun is probably the most an-cient and, at the same time, the most modern form of energy. Solar energy has been used for heating, cooking, and lighting since humans learned how to start a fire. Investment in solar energy technology in the United States, for example, came to $153.7 billion in 2017, up by 18% from the previous year. Excluding large hydropower (i.e., plants greater the 50 MW) this was by far the largest investment among all other types of renewable energy [1]. The sun radiates approximately 5.7×10^{24} joules of solar energy every year to the surface of the earth, of which about 30% is reflected back into space leaving about 4 million EJ available on Earth to be harnessed. This is more than 7,000 times the 560 EJ global primary energy consumption in 2017 [2]. Solar energy is used directly (e.g., for heating and lighting) and indirectly to generate electricity. Electricity is generated from solar energy using two approaches: solar thermal and solar electric. In solar thermal technology, mirrors and solar concentrators are used to collect and direct the sun's rays to special boilers where water is turned into steam to turn turbines that turn electric generators. In solar electric, photovol-taic modules are used to convert solar energy to electricity. This is by far the most popular and convenient technology. It involves less system components and no moving mechanical parts, and therefore requires relatively low mainte-nance. At the heart of any photovoltaic system is a simple semiconductor device called the photovoltaic cell, or simply the PV cell, which is typically a semicon-ductor pn-junction that converts the energy in sunlight to electricity. The con-struction, operation, and modeling of the PV cell is the subject of this chapter.

3.4   Electromagnetic Energy

Sunlight arriving on the surface of the earth is a form of electrical energy. To be more precise, it is called an electromagnetic energy, which travels as waves at the speed of light (i.e., $c = 3 \times 10^8$ m/s). An electromagnetic wave consists of a phenomenal number of very tiny discrete unit of energy called a photon. An electromagnetic wave, or a photon, is characterized by its frequency f(HZ) measured in hertz and wavelength λ(m) that are related by

$$f = c\lambda \tag{3.1}$$

A photon possesses an amount of energy, E measured in eV, given in terms of its frequency as

$$E = hf \tag{3.2}$$

Where h is Planck's constant and it is equal to 6.626×10^{-34} (J.s).

The spectrum of sunlight arriving on the earth's surface, when the sun is directly overhead (i.e., at the zenith), consists of 44% visible light, 3% ultraviolet, and the remainder is infrared. This amounts to approximately 1,000W per square meter; about 550 watts of infrared, 440 watts of visible light, and 30 watts of ultraviolet [3].

3.5   Spectral Irradiance, Insolation, and Radiation

Spectral irradiance I_λ is the common method used to characterize a light source. It is defined as the power received by a unit surface area in a wavelength differential $d\lambda$ and it is measured in W/m^2 μm. The insolation G, or the irradiance, is the integral of the spectral irradiance extended over all wavelengths of interest; that is

$$G = \int_{\lambda_1}^{\lambda_2} I_\lambda d\lambda \tag{3.3}$$

The unit for insolation is watts per square meter W/m^2.

The radiation is the time integral of the insolation extended over a given period, therefore it has the unit of energy (e.g.k Wh/m^2 day or k Wh/m^2 year) depending on the time slot used for the integration of the irradiance.

Example 3.1

A photovoltaic module that has an area of 0.3 m^2 receives a light whose spectral irradiance is 1000(W/m² μm) limited over the wavelength from $\lambda = 0.60$ to 0.7 μm. Calculate the insolation received and the radiation after one day.

Solution:

$$G = \int_{0.60\,\mu m}^{0.70\,\mu m} 1000 d\lambda = 100 \quad W/m^2$$

The radiation received after one day is

$$\text{Radiation} = a \int_{0}^{24h} G \cdot dt$$

$$= 0.3 \int_{0}^{24h} 100 \, dt = 720 \quad Wh.day$$

3.6   The Photovoltaic Effect

The photovoltaic effect refers to the generation of electric current by the energy contained in incident light waves, more specifically in the electromagnetic waves from the sun. We refer to any device or material that can convert the energy in incident electromagnetic photons into electricity as PV. The most common device, which is specifically manufactured to generate electricity when illuminated by sunlight, is the solar cell or simply the PV cell. Although the PV effect was discovered as early as 1839 by the French physicist Becquerel, the first true commercial PV cell did not appear until the late 1950s in the United States' artificial satellite the Vanguard I.

3.7   Photovoltaic Materials

Earlier PV cells used to be made from monocrystalline silicon. This is a very pure single-crystal silicon with continuous crystal lattice structure, which is relatively very expensive to prepare. Although less efficient, polycrystalline silicon that consist of grains of monocrystalline silicon has taken over because it is cheaper and easier to prepare. However, crystalline silicon cells tend to be relatively large with thickness of about 100 to 500 μm to maximize the absorption of incident photons. An alternative approach has been to manufacture thin film crystalline silicon on suitable substrate like glass [4]. Another crystalline material used for making PV cells is gallium arsenide (GaAs), which has good light absorption and has slightly wider bandgap than silicon. GaAs PV cells are more efficient than those made from silicon but are also a lot more expensive and they tend to be used only for demanding applications, such as spaceships and solar racing vehicles.

Another approach to making PV cell, other than crystalline wavers, is thin film technology using a variety of materials, such as amorphous silicon and cadmium telluride. Amorphous silicon is cheaper to prepare but at the cost of reduced efficiency. Further, amorphous silicon tends to degrade by approximately 20% after few months of use and then stabilizes [4]. Cadmium telluride (CdTe) is also relatively cheaper than crystalline silicon to prepare and has a wider bandgap than silicon.

Organic materials are also used for making PV cells; however, this is still the subject of research and reported efficiencies are generally low, less than 5% [5]. Current research is focused on new technologies, such as using multijunction PV devices in which different materials with different bandgaps are stacked on top of each other to maximize the wavelengths that can be absorbed [6]. We focus on silicon because silicon including poly- and monocrystalline still accounts to about 90% share of the crystalline silicon PV market [7].

3.8 The Bandgap Energy Theory

The atomic number of a chemical element is the number of protons in the nucleus of each atom of the element. This number is unique to every chemical element and since an atom is neutral, this number is the number of electrons in an atom. For example, the atomic number of silicon (Si) is 14, which means that there are 14 protons in the nucleus of the silicon and there are 14 electrons orbiting the nucleus of a silicon atom (see Figure 3.1). These 14 electrons are distributed into shells (or orbits) and a shell has a well-defined discrete energy level associated with it. An electron must exist in one of these levels and nowhere in between. A silicon atom has three shells with two electrons in the first shell (this is the first energy level), eight electrons in the second energy level, and finally four electrons in the third energy level. We say that electrons in silicon occupy three energy levels as shown in Figure 3.1 and each level is associated with a discrete amount of energy. These levels are also called energy states or bands. The highest energy band is the conduction band. The gaps between these energy states are called the forbidden energy states or forbidden bands. The last energy level before the conduction band is referred to as the valence band. Electrons in the valence band determine the electrical properties of the material, and hence, silicon is normally represented for simplicity as an atom of +4 charge on the nucleus surrounded by four orbiting electrons (the valence

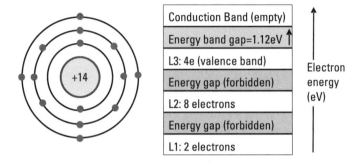

Figure 3.1 Energy levels in silicon.

electrons) as shown in Figure 3.2. The most significant energy gap is the one just before the conduction band and we refer to this as the bandgap energy E_g, which is measured in electron volt . For silicon, the bandgap energy is $E_g = 1.12$ eV.

An electron volt is defined as the amount of energy gained (or lost) by an electron in moving across a potential difference of one volt. One electron volt is the same as 1.602×10^{-19} joules.

An electron can move from one energy state to next higher one only if it is given an amount of energy that is at least equal to the bandgap energy between the two states. This means, in silicon, an electron must acquire an energy greater than or equal to 1.12 (eV) to enable it to leave the valence band and move into the conduction band and become a free electron available to contribute to current flow.

Electrons in the conduction band have the highest energy level and only those electrons can contribute to current flow. In metals, the conduction band is partially filled and, hence they (i.e., metals) can conduct electricity. In a semiconductor at absolute zero-degree temperature, the conduction band is empty, and hence, a semiconductor is a good insulator at zero-degree temperature. However, as the temperature of a semiconductor increases, some valence electrons can acquire sufficient thermal energy and breakaway from their parent

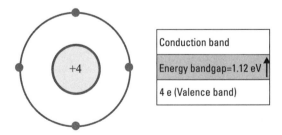

Figure 3.2 Energy band gap for silicon.

atoms and move up to the conduction band. This phenomenon is utilized in making thermistors, which are devices used for measurement of temperature. Similarly, when an electron moves form a higher energy state to a lower one, energy is released in the form of a photon (i.e., light wave). This phenomenon is used for making LEDs. In a silicon photovoltaic cell, when a photon from the sun with energy greater the bandgap energy of 1.12 (eV) falls onto the surface of the device, an electron can climb up to the conduction band as a free electron. However, when an electron leaves an atom, the atom left is an atom with one less negative charge, and hence, an unneutralized positive ion. This is equivalent to a positive charge carrier, which we model as a hole. It is called a hole because it represents an empty energy state, which a free electron can fall in it. In other words, the falling photon generated an electron-hole-pair (EHP). A PV cell is constructed so that an electric field sweeps these liberated electrons in one direction away from the holes to make a current; otherwise if these free electrons recombine with holes and an EHP is lost as a current currier, a photon is released as a result of the electron-hole recombination.

Example 3.2

What is the minimum frequency a photon must have to be able to create an electron-hole-pair in silicon? What is the corresponding maximum wavelength?
    Solution:
    For silicon $E_g = 1.12 eV$, using the relation $E = hf$,

$$f_{min} \geq \frac{E}{h} = \frac{1.12\ eV \times 1.6 \times 10^{-19}}{6.626 \times 10^{-34}}$$

$$f_{min} = 2.7 \times 10^{14}\ \text{Hz}$$

The maximum wavelength is obtained from $c = f\lambda$ as

$$\lambda_{max} = \frac{c}{f_{min}}$$

$$= \frac{3 \times 10^8}{2.7 \times 10^{14}}$$

$$\therefore \lambda_{max} = 1.11 \times 10^{-6}\ \text{or}\ 1.11\ \mu\text{m}$$

    Hence, for silicon PV cells, photons with frequency less than 2.7×10^{14} Hz or equivalently with wavelength longer than 1.11 μm have energy (hf) less than the bandgap energy for silicon of 1.12 (eV) and therefore cannot produce an electron-hole-pair that can contribute to current flow. Their energy is basically wasted in the silicon as heat. On the other hand, photons with wavelengths shorter than 1.11 μm have energy greater than 1.12 (eV) and therefore do excite electrons creating EHPs. However, the excess energy is again wasted as heat in the silicon structure. This fact about photons with less than or greater than the bandgap energy explains why there is a theoretical maximum limit on the efficiency of a solar cell as we will see in the next section.

3.9   Energy from the Sun and Its Spectrum

The spectrum of sunlight (i.e., the range of wavelengths it contains) travels through the atmosphere before reaching the surface of the earth. The atmosphere contains many particles and molecules of different sizes. Consequently, as sunlight passes through the atmosphere, its spectrum changes due to collisions, reflection, and absorption with those particles and molecules. The change to the spectrum depends on the length of the path through the atmosphere. The shortest path is when the sun is directly overhead (i.e., at the zenith); see Figure 3.3. As the sun moves closer to the horizon, sunlight passes through more atmosphere (i.e., the path length increases). The length of the path L_2 taken by the sunlight as it passes through the atmosphere, divided by the shortest possible path length L_1 that occurs when the sun is exactly overhead, defines the AM ratio as

$$AM = \frac{L_2}{L_1} = \frac{1}{\sin \alpha} \tag{3.4}$$

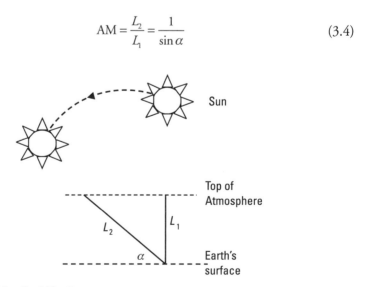

Figure 3.3   Defining the AM ratio.

Clearly, the AM ratio is a measure of how much atmosphere sunlight has to pass through before it reaches the earth's surface and consequently is a measure of how much of the light energy has been lost due to scattering and absorption. In other words, the AM ratio determines the terrestrial spectrum that has a direct impact on the responsivity of PV materials.

By convention AM0 means no atmosphere and AM1 means that the sun is directly overhead (i.e., at the zenith). Since the AM ratio is an indication of the spectrum and the responsivity of PV devices depends on the spectrum, it has been agreed to use a standard AM ratio of 1.5, which corresponds to sun making an angle of $\alpha = 42°$ above the horizon as a standard test condition for PV devices. The terrestrial spectrum at AM1.5 is shown in Figure 3.4, which indicates that the spectrum that reaches the surface of the earth ranges from approximately $0.3\,\mu\text{m}$ to about $3\,\mu\text{m}$[8]. That is, sunlight that reaches the surface of the earth contains photons of varying energy.

3.10   Theoretical Maximum Efficiency of PV Cells

We will now consider how the spectrum of the sun imposes a theoretical upper bound on the efficiency of a solar cell. We will consider silicon as an example and use the standard AM1.5 spectrum. For silicon the bandgap energy is 1.12 eV and we have seen that only photons with wavelengths shorter than $1.11\,\mu\text{m}$ have energy greater than 1.12 eV. Therefore, when a silicon PV cell is exposed to the standard AM1.5 spectrum, photons with longer wavelengths than $1.11\,\mu\text{m}$ will not be able to excite any electrons into the conduction band. Hence, their energy which amounts to approximately 20% of the total energy in the spectrum, is lost as heat in the silicon (see Figure 3.5). Further, photons with shorter wavelengths than $1.11\,\mu\text{m}$ have more energy than 1.12 eV and this excess energy is also lost, which is approximately 30% of the energy in the spectrum. This leaves approximately 50% of the energy in the AM1.5 spectrum usable for generating electron-hole pairs in silicon.

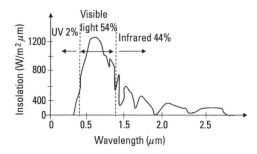

Figure 3.4   Terrestrial solar spectrum (AM1.5) [8].

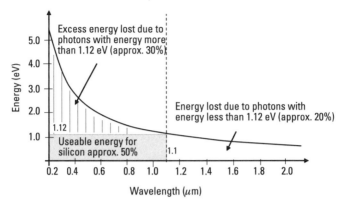

Figure 3.5   Energy of photons in the AM1.5 spectrum.

In fact, an upper limit on the efficiency of a silicon PV cell has been esti-mated using the AM1.5 solar spectrum as 49.6% [9]. That is, 50% is about the upper theoretical limit on the efficiency of silicon photovoltaic devices, which is imposed by the silicon bandgap. In practice, there are other factors that affect the efficiency of a solar cell, such as

- *Electron-hole recombination:* Light-generated electrons must be gener-ated close to the pn-junction so that they can diffuse to the junction prior to recombination. However, some of these electrons recombine with holes before diffusion, and hence, don't contribute to current.

- *Some photons are not absorbed in the cell.* For example, some photons are reflected off the top surface of the cell, while others may pass right through the cell or are blocked by the metal grid that collects current from the top of the cell.

- *Only about 50% to 70% of the full bandgap voltage appears across the terminals of PV cell.*

- *Power dissipation due to the internal effective resistance of the PV cell and due to the finite resistance of the metal conductors, which collect current from the cell.*

Considering all of the above factors, an upper bound on the efficiency of silicon PV cell has been estimated at about 31% at 1000 W/m² [10]. However, commercially available PV modules have efficiencies in the range of 12 to 20% [7].

Clearly, there is a trade-off between choosing a photovoltaic material that has a small bandgap against one with a large band gap. In a PV cell with a small-er bandgap, there will be more photos with energy sufficient to excite electrons into the conduction band, and hence, provide more light-generated current, but less terminal voltage. On the other hand, for a cell with a large bandgap, there will be less photons that can excite electrons; hence less current but higher terminal voltage. However, power is the product of current and voltage, hence, there must be an optimum bandgap, which should maximize the output power. This happens to be in the region of 1.2 to 1.8 (eV) [9]. The other possibility to alleviate the efficiency limitation due to the bandgap is to build multijunction PV devices so that each junction can absorb a certain range of wavelengths. For example, a PV device may include a top cell to absorb high-energy photons while permitting lower-energy photons to pass through to a lower cell with a smaller bandgap where they can be absorbed. However, this is currently costly and is an active area of research [6].

3.11   Operation of a Generic PV Cell

The structure of a silicon PV cell is similar to that of a pn-junction diode, as depicted in Figure 3.6, but with the p-type layer made thicker than the n-type layer. The p-type is grown over a metal substrate that makes the positive contact and then a thinner layer of n-type is grown. Finally, a metal grid is grown over the n-type to collect the liberated electrons and make the negative terminal. This is made from a grid to allow photons to pass through to the semiconductor.

When a photon with the correct amount of energy (i.e., greater than or equal to 1.12 eV) strikes the surface of the n-type, it can liberate an electron from the valence band and move it into the conduction band, thus generating an EHP in the n-type. Only those EHPs generated near the junction and within the diffusion length of minority carriers can contribute to current. When an EHP is generated in the n-region, electrons where they are majority carries are swept by the electrostatic field developed across the depletion region toward the

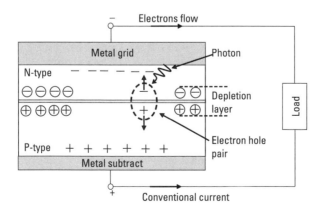

Figure 3.6   The structure of a generic PV cell is like that of a diode.

top side of the n-region (i.e., toward the negative contact). Generated holes that are the minority carriers in the n-region must be pushed away by the field across the junction into the p-region; otherwise a hole will recombine with an electron in the n-type leading to a loss of generated EHP and loss of current. Similarly, when EHPs are generated in the p-region, they generate minority electrons that must be swept away by the field to the top of the n-region before they are allowed to recombine with the majority holes. This is the reason the n-type is made thinner than the p-type region. Recombination of electrons and holes are considered as losses because they reduce the photon-generated current [11].

3.12   Terminal Characteristic of a PV Cell

There are two main curves that characterize a PV cell: the current-voltage (I-V) and power-voltage (P-V) characteristic. When a PV cell is completely covered so that no light can get to its surface and is excited by an external voltage source, V_{SS}, as shown in Figure 3.7 (a), the PV cell sinks current from the source and the dark current I-V characteristic, as show in Figure 3.7 (b), can be obtained. This is the I-V curve of the pn-junction diode that is described by the exponential equation, also known as Shockley's diode equation [12]:

$$I_d = I_o(e^{\frac{qV_d}{nkT}} - 1) \tag{3.5}$$

where

I_o = The reverse saturation current. This is the current due to the flow of minority carriers across the depletion region and is proportional to temperature. Typically, it is in the order of 10^{-10} to 10^{-12} A.

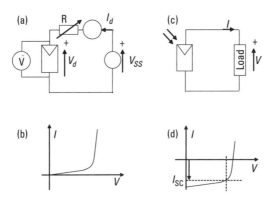

Figure 3.7   Obtaining the dark current characteristic of a PV cell, (b) the dark I-V curve, (c) illuminated PV cell supplying a load, and (d) the I-V curve for the illuminated cell.

V_d = The voltage drop across the cell. Typically, it is about 0.5V.

q = The electronic charge .

k = Boltzmann's constant, which is or joule per kelvin.

T = Temperature in kelvin.

n = Is the ideality factor. This is in the range of 1 to 2 for silicon.

When the external power supply is removed and the cell is illuminated, as shown in Figure 3.7(c), the current reverses (i.e., the cell is now sourcing current) and it appears as if the dark current characteristic has shifted down by the magnitude of the light-generated current, as shown in Figure 3.7(d). When an illuminated PV cell is supplying a load (i.e., sourcing current) it exhibits the terminal characteristic shown in Figure 3.8, which is obtained by flipping the characteristic in Figure 3.7(d) vertically. This I-V curve resembles the characteristic of a constant current source.

The output current of the cell is given by

$$I = I_{SC} - I_d \tag{3.6}$$

where I_{SC} is short-circuit light-generated current. Therefore, the equation describing the (I-V) characteristic becomes

$$I = I_{SC} - I_o(e^{\frac{qV_d}{nkT}} - 1) \tag{3.7}$$

The quantity KT/q is called the thermal equivalent voltage.

$$V_T = \frac{kT}{q} \tag{3.8}$$

Hence, we can rewrite the above current equation as

$$I = I_{SC} - I_o(e^{\frac{V_d}{nV_T}} - 1) \tag{3.9}$$

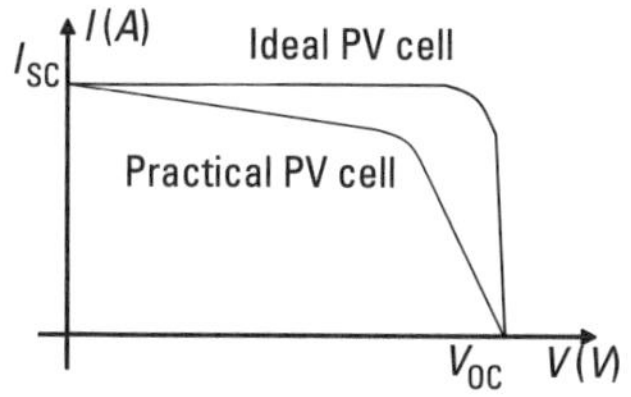

Figure 3.8 Ideal and practical terminal characteristic of a generic PV cell.

At room temperature of 25°C (i.e., 298° kelvin), the thermal voltage is approximately 25.7 mV. Hence, at room temperature and if $n = 1$, we can write

$$I = I_{SC} - I_o(e^{\frac{V_d}{0.0257}} - 1) = I_{SC} - I_o(e^{38.9V_d} - 1) \tag{3.10}$$

The short-circuit current varies almost linearly with insolation $G(W/m^2)$

$$I_{SC} = mG \tag{3.11}$$

Where m is a constant.

It is important to appreciate that this equation is an approximate model of the terminal characteristic of a PV cell. When the PV cell is open-circuited, we can solve for the open-circuit voltage (i.e., putting $I = 0$ in the above current equation):

$$0 = I_{SC} - I_o(e^{\frac{qV_{OC}}{nkT}} - 1)$$

$$I_{SC} = I_o(e^{\frac{qV_{OC}}{kT}} - 1)$$

$$\frac{I_{SC}}{I_o} + 1 = e^{\frac{qV_{OC}}{nkT}}$$

Take natural logs of both sides and rearranging

$$V_{OC} = \frac{nkT}{q} \ln\left(\frac{I_{SC}}{I_o} + 1\right) \tag{3.12}$$

At room temperature (i.e., 298K) and $n = 1$, this simplifies to

$$V_{OC} = 0.0257 \ln\left(\frac{I_{SC}}{I_o} + 1\right) \tag{3.13}$$

The open-circuit voltage is almost independent of the level of insolation but changes considerably with temperature. For a silicon PV cell, the open-circuit voltage decreases by about 0.37% per degree rise in temperature [9].

The P-V curve is obtained from the I-V curve through point by point multiplication of the current and voltage curve ($P = VI$), as shown in Figure 3.9.

We can identify one special point on the P-V curve for which the power reaches its maximum value (i.e., the point (I_{mp}, V_{mp})). At this point P_{max} known as the maximum power point (MPP), the PV cell generates its maximum power

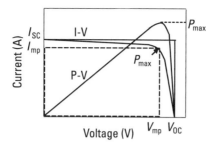

Figure 3.9 Terminal characteristics of a PV cell under given level of irradiance and temperature.

and therefore, ideally, a cell must always be operated at this point. The MPP is always located near the bend of the I-V curve. Note that the maximum power is represented by the largest rectangle that can be fitted under the I-V curve, which is shown by the dashed lines:

$$P_{max} = I_{mp}V_{mp} \tag{3.14}$$

The open-circuit voltage V_{OC} and short-circuit current I_{SC} are the maximum voltage and maximum current, respectively, and their product is represented by the larger rectangle. However, the power at either of these two points is clearly zero. The fill factor (FF) is a figure of merit of a PV cell, which determines the maximum power from a solar cell using these two values and is defined as

$$FF = \frac{V_{mp}I_{mp}}{V_{OC}I_{SC}} \tag{3.15}$$

The FF may be expressed as a decimal fraction or a percentage. The FF is seen as the ratio of the area of the smaller ($V_{mp} \times I_{mp}$) rectangle to the area of the larger rectangle ($V_{OC} \times I_{SC}$). The FF can also be seen as measure of the squareness of the I-V curve and it is typically in the range of 0.6 to 0.8.

If the active surface area of a PV is $a(m^2)$ and the incident insolation is $G(W/m^2)$, then the input power P_{in} (W) to the cell is

$$P_{in} = aG \quad (W) \tag{3.16}$$

The efficiency of the cell is defined as

$$\eta = \frac{P_{max}}{P_{in}} \times 100\% \tag{3.17}$$

3.13 Standard Test Conditions

Since the characteristics of a PV cell change with temperature and irradiance, standard test conditions (STCs) have been agreed on in order to enable fair comparisons between different PV cells made by different manufacturers. Those test conditions include solar insolation of 1000 W/m^2, which is also known as 1 Sun, with spectral distribution at AM1.5 and room temperature of 25°C.

Example 3.3

A solar cell has a reverse saturation current of $I_o = 10^{-10}$ A and a short-circuit current of 5A at full insolation and room temperature. Given that its ideality factor is unity, determine its open-circuit voltage at full insolation and the percentage change in the open-circuit voltage if the insolation is reduced by 50%.
  Solution:
  At full insolation, the open-circuit voltage is

$$V_{\text{OC}} = V_{\text{T}} \ln(\frac{I_{\text{SC}}}{I_o} + 1) = 0.0257 \ln(\frac{5}{10^{-10}} + 1) = 0.6331\,V$$

  At 50% insolation, the short-circuit current is reduced by 50% (i.e., I_{SC} = 2.5 A)

$$V_{\text{OC}} = V_{\text{T}} \ln(\frac{I_{\text{SC}}}{I_o} + 1) = 0.025 \ln(\frac{2.5}{10^{-10}} + 1) = 0.6153\,V$$

The percentage drop in the open-circuit voltage is therefore

$$V_{\text{OC}} = \frac{0.6331 - 0.6153}{0.6331} \times 100\% = 2.8\%$$

That is, for a 50% change in insolation, the open-circuit voltage only changed by 2.8%, but the short-circuit current is halved.

3.14 Equivalent Circuit Models of a PV Cell

We will now develop a number of equivalent circuit models for a PV cell. The first two important parameters of a PV cell are the short-circuit current (I_{SC}) and the open-circuit voltage V_{OC}. The short-circuit current is the current that flows through the output terminals of the PV when shorted, while the open-circuit voltage is the voltage measured across those terminals when open-circuited. Since a PV cell is basically a constant current source, let us define some terminologies related to constant current and constant voltage sources. A voltage source is rated by its open-circuit voltage (i.e., its electromotive force E(V)) and its full-load current. An ideal voltage source should maintain its terminal voltage; that is, the voltage across its load is constant over the full range of its rated current as suggested in Figure 3.10.

However, in a practical voltage source, as the load increases, the terminal voltage decreases, as shown in Figure 3.10(b). This means that the voltage source has some internal resistance that is represented by $r(\Omega)$ in the equivalent circuit of Figure 3.10(c). The terminal voltage is given by

$$V_L = E - Ir \tag{3.18}$$

Clearly, the internal resistance $r(\Omega)$ is zero for an ideal voltage source and should be as small as possible for a practical source.

An ideal constant current source should supply its rated short-circuit current I_{SC} over a range of terminal voltages from zero to its full rated voltage, as shown in Figure 3.11. That is, the load current I_L should be the same as the short-circuit current I_{SC} over the full range of voltage. However, for a practical current source, the load current decreases as the load resistance increase. This fact can be modeled by shunt (i.e., parallel resistance across the current source). Therefore, the load current is given by

$$I_L = I_{SC} \frac{R_p}{R_p + R} \tag{3.19}$$

Clearly, the internal resistance R_p (Ω) is infinite for an ideal current source and should be as large as possible for a practical source so that most of the source current flows through the load.

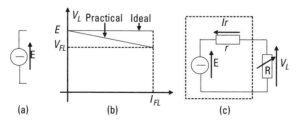

Figure 3.10 (a) Constant voltage source symbol, (b) ideal and practical terminal characteristics, and (c) equivalent circuit of a practical source.

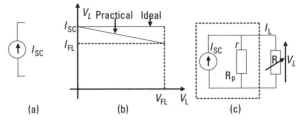

Figure 3.11 (a) Constant current source symbol, (b) ideal and practical terminal characteristics, and (c) equivalent circuit of a practical source.

3.14.1 Ideal Model

The ideal model is the simplest equivalent circuit model of a PV cell. It consists of a constant current source in parallel with a diode, as shown in Figure 3.12. This ideal current source (I_{SC}) supplies current to the load and to the diode. Hence, the terminal load current is given by

$$I = I_{SC} - I_o(e^{\frac{qV}{nkT}} - 1) \tag{3.20}$$

Example 3.4

A solar cell has a reverse saturation current of $I_o = 5 \times 10^{-10}$ A, an ideality factor of 1, a short-circuit current of 4A, and an open-circuit voltage of 0.5V at standard test conditions. Use the ideal model of a PV cell to predict the load current.

$$I = I_{SC} - I_o(e^{\frac{V}{0.0257}} - 1) = 4 - 5 \times 10^{-10}(e^{\frac{0.5}{0.0257}} - 1) = 3.860 \ \text{A}$$

3.14.2 Ideal Model Plus Parallel Resistance Model

The above model (i.e., the ideal model) is not accurate enough and can fail easily. For example, consider the situation when two PV cells are connected in

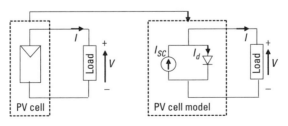

Figure 3.12 The ideal PV cell model.

series, as shown in Figure 3.13, with the top cell completely covered so that no light can get to its surface and hence it cannot produce any current. Since the top cell is not producing any current and its diode will be reverse biased by the voltage from the bottom illuminated cell, the ideal model tells us that no current can flow through it and therefore, no power will be delivered to the load.

In practice, there will be power delivered to the load, although reduced due to shading. Therefore, the simple ideal model fails in a situation like the one described above and a better model is needed. Since current is still flowing through the shaded cell, this suggests the presence of a parallel resistance across the cell, as shown in Figure 3.14.

The load current is now modified to account for the parallel resistance as

$$I = \left(I_{\text{SC}} - I_{\text{d}}\right) - I_{\text{P}} \tag{3.21}$$

where

$$I_{\text{P}} = \frac{V}{R_{\text{P}}} \tag{3.22}$$

The load current is therefore

$$I = I_{\text{SC}} - I_{\text{o}}(e^{\frac{qV}{nkT}} - 1) - \frac{V}{R_{\text{P}}} \tag{3.23}$$

Therefore, the above equation tells us that for any given voltage (V), the parallel (also known as the leakage resistance) reduces the load current below the value predicted by the simple model by the quantity (V/R_{p}), as illustrated in Figure 3.15.

Note that if the losses due to the parallel resistor are to be less than say 1%, then the parallel resistance must satisfy the inequality

$$R_{\text{P}} > \frac{100 \, V_{\text{OC}}}{I_{\text{SC}}} \tag{3.24}$$

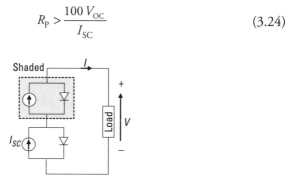

Figure 3.13   Two PV cells connected in series with the top one completely shaded.

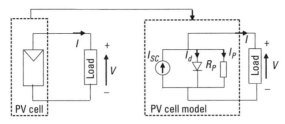

Figure 3.14　Ideal plus shunt resistance model.

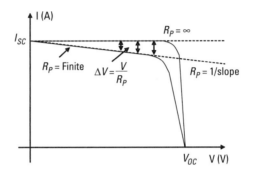

Figure 3.15　Effect of the parallel resistance on the cell output current.

As an example, for a cell rated at $I_{SC}= 5A$ and $V_{OC}= 0.5V$, the parallel resistor must be greater than 10Ω.

Example 3.5

A solar cell has a reverse saturation current of $I_o = 5 \times 10^{-10}$ A, an ideality factor of 1, a short-circuit current of 4A, an open-circuit voltage of 0.5V at standard test conditions, and a parallel effective resistance of 10 ohms. Use the ideal model plus parallel resistance model of a PV cell to determine the load current.

$$I = I_{SC} - I_o(e^{\frac{V}{0.0257}} - 1) - \frac{V}{R_P} = 4 - 5\times 10^{-10}(e^{\frac{0.5}{0.0257}} - 1) - \frac{0.5}{10} = 3.81 \text{ A}$$

3.14.3　Ideal Model Plus a Series Resistance

In any real PV cell, current flow through the cell experiences some resistance due to the metal contacts with the semiconductor and due to its flow through the semiconductor material itself. To account for this resistance a series resistance R_S is included in the simple model, as shown in Figure 3.16. Now we analyze this equivalent circuit as follows.

Applying KVL, we can easily see that the junction voltage V_d is now the sum of the load voltage, that is, the cell's terminal voltage and the voltage drop across the series resistance

$$V_d = V_S + V \tag{3.25}$$

where

$$V_S = IR_S \tag{3.26}$$

The load current is therefore given by

$$I = I_{SC} - I_o(e^{\frac{q(IR_S+V)}{nkT}} - 1) \tag{3.27}$$

The terminal voltage of the cell is $V = V_d - V_S$, which can be interpreted as the ideal I-V curve but with the voltage at any value of the current shifted to the left (i.e., decreased) by the amount of the voltage drop across the series resistance (i.e., by the amount $\Delta V = IR_S$), as shown in Figure 3.17.

For a PV cell to have less than 1% losses due to the series resistance, the series resistance must satisfy the inequality

$$R_S < \frac{0.01V_{OC}}{I_S} \tag{3.28}$$

For a 5A 0.5V cell, this means that

$$R_S < \frac{0.01(0.5)}{5} = 0.001 \ \Omega$$

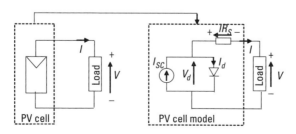

Figure 3.16   Ideal model plus series resistance model.

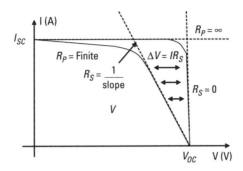

Figure 3.17   Effect of the series resistance on the original I-V curve.

3.14.4   Ideal Model Plus Parallel Resistance Plus Series Resistance

Clearly, a more accurate model is the model that includes the effects of both the parallel and the series resistances, as shown in Figure 3.18.

Applying Kirchhoff's current law at the node above the diode's anode, the output current of the cell can be written as

$$I = I_{SC} - I_d - I_P \tag{3.29}$$

The voltage across the diode (i.e., the junction voltage) is given by

$$V_d = V + IR_S \tag{3.30}$$

The current through the parallel resistor is

$$I_P = \frac{V + IR_S}{R_P} \tag{3.31}$$

The load current is therefore given as

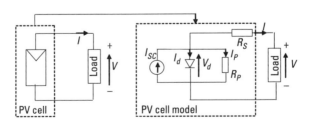

Figure 3.18   PV model with both series and parallel resistances.

$$I = I_{SC} - I_o(e^{\frac{q(V+IR_S)}{nkT}} - 1) - \left(\frac{V + IR_S}{R_p}\right) \qquad (3.32)$$

At room temperature and assuming the ideality factor is unity, this equation can be written as

$$I = I_{SC} - I_o(e^{\frac{V_d}{0.0257}} - 1) - \frac{V_d}{R_p} \qquad (3.33)$$

Clearly, this is an implicit equation because V and I appear on both sides of the equation and therefore, an explicit solution for either V or I is not readily possible. One technique of expressing the voltage as an explicit function of the current uses the Lambert W function, which is beyond the scope of this book [13]. Alternatively, a spreadsheet method may be used. In this method, we start with an initial value of the junction voltage V_d and use it to calculate the current. The cell's voltage is then determined from $V = V_d - IR_S$. The value of the diode voltage V_d is then incremented in steps, and at each step the above process of calculating I and V is repeated. This assumes that values of the series and parallel resistors are known. However, those are not normally available and must be extracted either analytically or experimentally [14].

The above models of a PV cell are based on the single-diode model due to its simplicity and sufficient accuracy for many practical situations [15]. However, there are also models based on a double-diode equivalent circuit. The extra diode accounts for the recombination loss in the PV cell [16].

Example 3.6

A solar cell has a reverse saturation current of $I_o = 5 \times 10^{-10}$A, a unity ideality factor and a short-circuit current of 8A at standard test conditions. The effective parallel and series resistances are 10 and 0.02 ohms, respectively. Use a Microsoft Excel spreadsheet or equivalent, together with the ideal model plus parallel and series resistances equivalent circuit of a PV cell to predict the load current.

The current equation of the cell is

$$I = I_{SC} - I_d - I_P$$

At STC:

$$I_d = I_o(e^{\frac{V_d}{0.0257}} - 1), \quad I_P = \frac{V_d}{R_p}$$

Given

$$R_S = 0.02\ \Omega,\ I_o = 5 \times 10^{-10}\ \text{A},\ I_{SC} = 8\ \text{A}$$

We start with an initial value of the voltage, say $V_d = 0.4\text{V}$, the current is $I = 4 - I_o(e^{\frac{0.5}{0.0257}} - 1) - \dfrac{0.4}{R_p} = 7.957$. This value of current is then used to find the voltage across the cell as

$$
\begin{aligned}
V = V_d - IR_S \\
= 0.4 - 0.02 \times 7.957 \\
= 0.241\ \text{V}
\end{aligned}
$$

The junction voltage is then incremented by, say 0.2V, and the process is repeated. The spreadsheet is shown in Figure 3.19 along with a plot of the I-V curve.

3.15   Summary

Photovoltaic technology is the most direct method of converting the energy in the sunlight into electricity. Electricity is the most versatile form of energy and solar energy is the most abundant source of renewable energy. PV cells are made from semiconductor material, mainly silicon, in the form of a pn-junction. Incident light generates electrons and holes that are separated by the electrostatic field created across the junction to create a terminal voltage. This voltage circulates a current when the cell is connected in a circuit. The spectrum of sunlight contains photons of varying energy that imposes an upper limit on the efficiency of semiconductor PV cells.

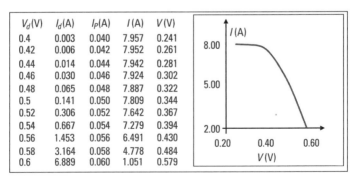

V_d(V)	I_d(A)	I_p(A)	I(A)	V(V)
0.4	0.003	0.040	7.957	0.241
0.42	0.006	0.042	7.952	0.261
0.44	0.014	0.044	7.942	0.281
0.46	0.030	0.046	7.924	0.302
0.48	0.065	0.048	7.887	0.322
0.5	0.141	0.050	7.809	0.344
0.52	0.306	0.052	7.642	0.367
0.54	0.667	0.054	7.279	0.394
0.56	1.453	0.056	6.491	0.430
0.58	3.164	0.058	4.778	0.484
0.6	6.889	0.060	1.051	0.579

Figure 3.19   Spreadsheet and the I-V plot.

3.16 Problems

P3.1   a. Explain the terms spectral irradiance, irradiance, and insolation.

b. Explain the bandgap theory in relation to electrical conductivity of metals.

c. Explain the photovoltaic effect.

d. Define the air mass ratio and explain its significance to the photovoltaic effect.

e. If a PV cell is made to operate in space, it must be made from different material to that made to operate on Earth. Explain why.

f. Explain the significance of the bandgap energy in relation to the photovoltaic effect.

g. Define the fill factor and explain its significance.

h. Explain why there is an upper limit on the efficiency of a silicon PV cell.

P3.2   Determine the minimum frequency and the corresponding maximum wavelength of electromagnetic photons capable of generating electron-hole-pairs in the following materials:

a. Cadmium telluride with a bandgap of 1.44 eV.

b. Gallium arsenide with a bandgap of 1.42 eV.

c. Germanium with a bandgap of 0.66 eV.

P3.3   At room temperature, a silicon PV cell has a dark current of 200 mA and a voltage drop is 0.6V. Assuming the ideality factor is 1, determine it reverse saturation current.

P3.4   A silicon solar cell supplying a resistive load has a reverse saturation current of $I_o = 10^{-10}$A and a short-circuit current of 5A at standard test conditions. Making any reasonable assumptions and given that its ideality factor is unity

a. Determine its open-circuit voltage.

b. Determine the load current when the load voltage is 0.5V.

c. Determine the power delivered to the load.

d. Given that the active area of the cell is 0.01m², determine its efficiency.

P3.5   At standard test conditions, a silicon solar cell has a reverse saturation current of $I_o = 10^{-10}$A, a short-circuit current of 2A, and an effective parallel resistance of 7.5 ohms. When the cell is connected across a resistive load, its output voltage is 0.58V. Making any reasonable assumptions and given that its ideality factor is unity

a. Determine its open-circuit voltage of the cell at full insolation.

b. Determine the load current.

c. Determine the power delivered to the load.

d. Determine the efficiency of the cell given that its active area is 0.006 m^2

P3.6   A silicon solar cell has a reverse saturation current of $I_o = 5 \times 10^{-10}$A and a short-circuit current of 8 A at standard test conditions. It has an effective parallel resistance of 10 ohms and a series resistance of 0.01 ohms. Making any reasonable assumptions and given that its ideality factor is unity

a. Determine the output voltage, current, and power delivered by the cell assuming the junction voltage is 0.5V.

b. Use spreadsheet to determine the cell's current and voltage as the junction voltage is increased from 0.5 to 0.6V in steps of 0.1V and plot the I-V and P-V characteristics.

P3.7   A silicon PV cell whose ideality factor is unity has the I-V characteristic shown in Figure 3.20 obtained under standard test conditions. Making any reasonable assumptions

a. Estimate the equivalent parameters of the ideal model plus series and parallel resistances of this cell.

b. Assuming that the active area of the cell is 0.012 m^2, estimate its efficiency.

c. Sketch the P-V characteristic and estimate the fill factor of the cell.

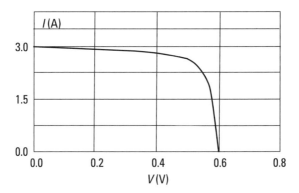

Figure 3.20   I-V characteristic for Problem P.7

References

[1]   Frankfurt School of Finance and Management, "Global Trends in Renewable Energy Investment 2018," FS-UNEP Collaborating Centre for Climate and Sustainable Energy Finance, Frankfurt, 2018.

[2]   BP, "Statistical Review of World Energy," BP, 2018, https://www.bp.com/content/dam/bp/business-sites/en/global/corporate/pdfs/energy-economics/statistical-review/bp-stats-review-2018-full-report.pdf.

[3]   The University of Tennessee Institute of Agriculture, "Solar and Sustainable Energy," 2018, https://ag.tennessee.edu/solar/Pages/What%20Is%20Solar%20Energy/Sunlight.aspx.

[4]   Green, M., K. Emery, Y. Hishikawa, W. Warta, and E. Dunlop, "Solar Cell Efficiency Table," *Progress in Photovoltaics*, Vol. 19, No. 5, 2011, pp. 565–572.

[5]   Xue, J., S. Uchida, B. P. Rand, and S. R. Forrest, "Asymmetric Tandem Organic Photovoltaic Cells with Hybrid Planar-Mixed Molecular Heterojunctions," *Applied Physics Letters*, Vol. 85, 2015, p. 5757.

[6]   Cariou, R., K. Medjoubi, L. Vauche, et al., "Evaluation of III-V/Si Multi-Junction Solar Cells Potential for Space," *2018 IEEE 7th World Conference on Photovoltaic Energy Conversion*, Waikoloa Village, 2018.

[7]   Glunz S. W., and D. Biro, "Crystalline Silicon Solar Cells: State-of-the-Art and Future Developments," *Comprehensive Renewable Energy*, Vol. 1, 2012, pp. 353–387.

[8]   NASA, "Terrestrial Photovoltaic Measurement Procedures," NASA/1022-77/16 NASA TM 73702, Cleveland, OH, 1977.

[9]   Masters, G. M., *Renewable and Efficient Electric Power Systems*, New Jersey: John Wiley & Sons Inc., 2004.

[10] Henry, C. H., "Limiting Efficiencies of Ideal Single and Multiple Energy Gap Terrestrial Solar Cells," *Journal of Applied Physics*, Vol. 51, No. 8, 1980, p. 4494.

[11] Khaligh, A., and O. C. Onar, *Energy Harvesting Solar, Wind, and Ocean Energy Conversion Systems*, New York: CRC Press, 2010.

[12] Shockley, W., "The Theory of P-N Junction in Semiconductors and P-N Junction Transistors," *The Bell System Technical Journal*, Vol. 28, No. 3, 1949, p. 454.

[13] Batzelis, E. I., I. A. Routsolias, and S. A. Papathanassiou, "An Explicit PV String Model Based on the Lambert W Function and Simplified MPP Expressions for Operation Under Partial Shading," *IEEE Transactions on Sustainable Energy*, Vol. 5, No. 1, 2014, pp. 301–312.

[14] Ibrahim, H., and N. Anani, "Evaluation of Analytical Methods for Parameter Extraction of PV modules," *Energy Procedia*, Vol. 134, 2017, pp. 69–78.

[15] Weidong, X., W. G. Dunford, and A. Capel, "A Novel Modeling Method for Photovoltaic Cells," *2004 IEEE 35th Annual Power Electronics Specialists Conference*, Aachen, 2004.

[16] Haque, K., Z. Salam, and H. Taheri, "Solar Energy Materials and Solar Cells," *Solar Energy Materials and Solar Cells*, Vol. 95, No. 2, 2011, pp. 586–594.

4

Photovoltaics: PV Arrays Operation and Characteristics

4.1 Learning Outcomes

After actively engaging with the material in this chapter, you should be able to

1. Describe the need to construct PV arrays;
2. Explain the effects on the I-V and P-V characteristics of PV modules when connected in series and parallel;
3. Perform simple voltage, current, and power calculations on PV arrays;
4. Explain the effects of temperature on the characteristics of PV modules;
5. Explain the effects of varying insolation on the characteristics of a PV module;
6. Explain and assess the effects of partial shading on the I-V characteristics of PV modules;
7. Explain the need for bypass diodes in PV modules;
8. Explain the need for blocking diodes in PV arrays.

4.2   Overview

In this chapter, we will study the construction of photovoltaic modules and arrays and their terminal characteristics. We will also consider the effects of changing insolation and temperature on the characteristics of PV cells and modules. The adverse effects of partial shading on the performance of PV systems is also discussed along with methods of mitigating these effects. Finally, several problems at the end of the chapter are provided to consolidate understanding of the material presented in this chapter.

4.3   Introduction

A single photovoltaic cell can generate only a small amount of power, typically around 0.5V and 4A under full illumination, which makes a single PV cell impracticable for most applications. Therefore, it is the norm that several cells are connected electrically in series and manufactured in a single package known as a PV module, or panel, to provide higher voltage and power. A PV module consists of individual PV cells, typically 36, 60, or 72 connected in series. In practice, a 36-cell module is normally designated as a 12V module although the voltage may be much higher than that, typically in the range of 18 to 22V. For example, the 36-cell Mitsubishi PV-MF110EC3 PV module has an open-circuit voltage of 21.2V and a maximum power of 110W under standard test conditions [1]. A 72 cell module designated as a 36-volt module typically has an open-circuit voltage in the region of 45V and a maximum power of about 300W.

4.4   Photovoltaic Arrays

Photovoltaic modules are prewired during the manufacturing process and are made in environmentally protective sealed packages. They are reliable products with a lifespan in excess of 20 years. PV modules are rated by their maximum (or peak) DC output power, open-circuit voltage, and short-circuit current under STC. The STCs are defined by the module operating at temperature of 25°C and insolation of 1,000W per square meter (or 1 Sun) under air mass ratio AM1.5 spectrum. However, in practice PV modules operate at much higher temperatures, and therefore, their performance falls below their STC rating. For most practical applications, one or two modules cannot provide a sufficient amount of power, and therefore, PV modules are connected in series or in parallel or as a combination of series and parallel connections to make a PV array, as shown Figure 4.1.

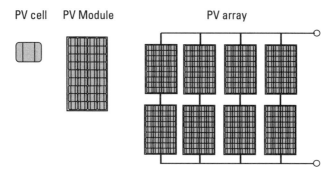

Figure 4.1 PV cell, module, and array.

The actual configuration (i.e., how many modules are in series and/or in parallel) in a PV array rests with the PV system designer. This depends on the required output voltage and current from the array. When we connect PV cells, or module, in series, the same current flows in each one of them and the total voltage is the sum of the individual voltages. Consider for example, 36 cells connected in series in one module as shown in Figure 4.2. If each cell has an open-circuit voltage of 0.5V, the open-circuit voltage of the module will be 18V. The terminal I-V characteristic of the module is obtained by adding the voltages of the individual cells, while the short-circuit current is the same for the module as it is for an individual cell as illustrated in Figure 4.2.

We have seen in the previous chapter that the voltage across one PV cell is given by $(V_d - IR_S)$ where R_S is the internal equivalent series resistance and V_d is the junction voltage of one PV cell. The voltage across the whole module is

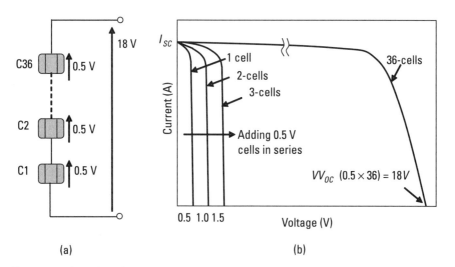

Figure 4.2 A 36 PV cell module.

the voltage across one cell multiplied by the number of cells in series. Hence, if there are n cells, the total module voltage V_{mod} is

$$V_{\text{mod}} = n(V_d - IR_S) \tag{4.1}$$

When identical modules are connected in parallel, they all share the same voltage across them, but the total current in the load is the sum of the individual module currents, as shown in Figure 4.3, which also shows the I-V curve of the combination.

Example 4.1

Given a number Z of identical PV modules, show that whether they are connected in series or in parallel, or as any combination of series and parallel branches, the output power is always the same for a given insolation.

Example 4.2

A PV module consists of 36 identical cells connected in series. Under standard test conditions, each cell has the following particulars:

$$I_{\text{SC}} = 5 \text{ A}, \quad I_o = 5 \times 10^{-10} \text{ A}, \quad R_S = 0.02 \ \Omega, \quad R_P = 7 \ \Omega$$

Determine:
   a. The voltage, current, and power delivered by the module when each
       cell has a junction voltage of 0.5V;

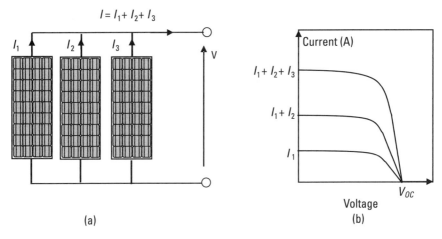

(a)   (b)

Figure 4.3   Connecting three PV modules in parallel.

b. Use Microsoft Excel or a similar program to develop a spreadsheet to calculate the module's current, voltage, and power for a range of cell voltages from 0.5V to 0.58V in steps of 0.1V.

Solution:

Using the equivalent circuit model of the PV cell we studied in the previous chapter (shown in Figure 4.4), the module current is given by

$$I = I_{SC} - I_{d} - I_{P} \tag{4.2}$$

where

$$I_{d} = I_{o}(e^{38.9V_{d}} - 1) \tag{4.3}$$

and

$$I_{P} = \frac{V_{d}}{R_{P}} \tag{4.4}$$

Therefore, the module's current is given by

$$I = 5 - 5 \times 10^{-6}(e^{38.9(0.5)} - 1) - \frac{0.5\,V}{7\,\Omega} = 4.788\ \text{A}$$

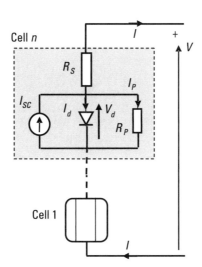

Figure 4.4   A PV module with n cells in series and with the nth cell represented by its equivalent circuit model.

The module's voltage is

$$V_{mod} = 36(0.5 - 5(0.02)) = 14.533\ V$$

Power delivered is

$$P_{mod} = V_{mod}I = 14.553\ V \times 5\ A = 69.677\ W$$

The spreadsheet is shown in Figure 4.5. We started with an initial value of the junction voltage $V_d = 0.5$ V and incremented it in steps of 0.1V.

Note that for junction voltages greater than about 0.59, we start getting negative values for the module current. This is because the sum of the diode and the parallel resistance currents become greater than the short-circuit current, which is not realistic. An added advantage of the spreadsheet method is the fact that it allows us to plot the I-V and P-V characteristics as shown in Figure 4.6.

4.5 Fill Factor and Efficiency

The efficiency and the FF for a PV module are defined in the same way as in the previous chapter for a single PV cell. Figure 4.7 shows the I-V and P-V curves for a generic PV module. The maximum power point labeled MPP is the point at which the power delivered by the PV module is a global maximum. The voltage and current corresponding to this point are denoted as V_{mp} and I_{mp}, respectively. These define the maximum area of a rectangle that can be fitted under the I-V curve (shown by the dashed lines), which is the rated (maximum) power of the module

V_d(V)	I_d (A)	I_P (A)	I (A)	V (V)	V_{mod}(V)	P_{mod}(A)
0.5	0.14	0.07	4.788	0.404	14.553	69.677
0.51	0.21	0.07	4.720	0.416	14.962	70.613
0.52	0.31	0.07	4.619	0.428	15.394	71.110
0.53	0.45	0.08	4.472	0.441	15.860	70.929
0.54	0.67	0.08	4.256	0.455	16.376	69.691
0.55	0.98	0.08	3.937	0.471	16.965	66.792
0.56	1.45	0.08	3.467	0.491	17.664	61.244

Figure 4.5 Spreadsheet for Example 4.2.

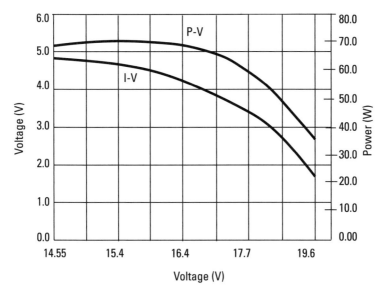

Figure 4.6 I-V and P-V curves for Example 4.2.

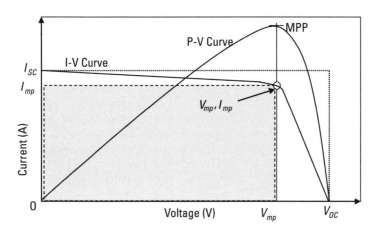

Figure 4.7 I-V and P-V curves of a typical PV module.

$$P_{\max} = V_{\mathrm{mp}} I_{\mathrm{mp}} \tag{4.5}$$

The product of the short-circuit current and the open-circuit voltage (i.e., $I_{SC} \times V_{OC}$) define the larger rectangular area shown by the dotted line. These two rectangles define the fill factor as the ratio of the power at the maximum power point (i.e., the area of the smaller rectangle) to area of the larger rectangle

$$FF = \frac{I_{mp}V_{mp}}{I_{sc}V_{OC}} \qquad (4.6)$$

This may be expressed as a decimal fraction or as a percentage. The fill factor typically varies between about 50% to 80% and clearly the higher the FF the better is the module.

If the active surface area of a PV module is a(m^2) and the incident insolation is G(W/m^2), then the input power P_{in}(W) to the module is

$$P_{in} = aG \qquad (4.7)$$

The efficiency η of the module is defined as the ratio of the maximum output power to the input power and is normally expressed as a percentage as

$$\eta = \frac{P_{max}}{P_{in}} \times 100\% \qquad (4.8)$$

4.6   Effects of Temperature and Insolation on the I-V and P-V Characteristics

The performance of a PV module varies with temperature and insolation. This variation is manifested in the variations of the I-V and P-V characteristic curves that are normally shown on the manufactures' datasheets. Typical variation of the short-circuit current and open-circuit voltage of a PV module are shown in Figure 4.8.

The short-circuit current is directly proportional to the insolation G(W/m^2)

$$I_{SC} = mG \qquad (4.9)$$

where m is a constant. That is, doubling the insolation doubles the short-circuit current and vice versa. The open-circuit voltage, however, varies with insolation on a logarithmic scale and for most practical purposes the percentage drop in the voltage is negligible compared to the decrease in the short-circuit current for the same decrease in insolation level.

However, the open-circuit voltage is sensitive to variations in temperature and as the temperature increases, the voltage decreases considerably, while the short-circuit current increases by a very small amount compared to the decrease in the open-circuit voltage for the same increase in temperature. For crystalline

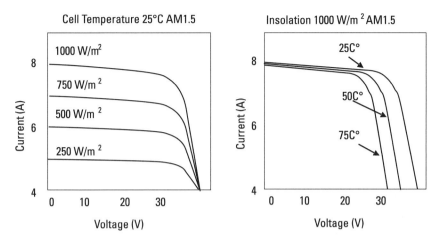

Figure 4.8 Typical current-voltage characteristic curves under various temperatures and insolation levels.

PV silicon, the loss in the open-circuit voltage is about 0.37% per degree rise in temperature, and we define the open-circuit voltage temperature coefficient as $\alpha_V = 0.0037/°C$. Therefore, given that V_{OCT1} is the open-circuit voltage at temperature T_1, which is the STC temperature (i.e. $T_1 = 25°C$)) the open-circuit voltage at any other temperature T_2 is given by

$$V_{OCT2} = V_{OCT1}[1 - \alpha_V(T_2 - T_1)] \tag{4.10}$$

The increase in the short-circuit current for crystalline PV silicon is very small—about 0.05% per degree rise in temperature—and we define the short-circuit current temperature coefficient as $\alpha_I = 0.0005/°C$. Therefore, given that I_{SCT1} is the short-circuit current at temperature $T_1 = 25°C$, the short-circuit current at any temperature T_2 is given by

$$I_{SCT2} = I_{SCT1}[1 + \alpha_I(T_2 - T_1)] \tag{4.11}$$

The combined effects of the temperature rise on the voltage and current of a PV cell is that that the MP moves slightly up and to the left, as shown in Figure 4.9, with a net drop in the maximum power of about 0.5% per degree rise in temperature.

Therefore, given that the power temperature coefficient is $\alpha_P = 0.005/°C$ and P_{maxT1} is the maximum power at temperature T_1, the maximum power at any temperature T_2 is given by

$$P_{maxT2} = P_{maxT1}[1 - \alpha_P(T_2 - T_1)] \tag{4.12}$$

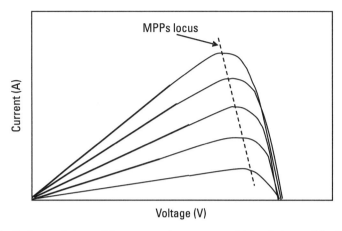

Figure 4.9 Typical variations of the power-voltage curves with insolation and the MPP locus.

Clearly, when a PV module is connected to a load, ideally the operating point should be kept at the MPP. However, since this point changes with insolation and temperature, in practice operation at the MPP is maintained using special electronic device called a maximum power point tracker (MPPT).

PV modules are rated at STC which means at temperature of 25°C and insolation of 1,000(W/m^2). However, most of the time, PV modules operate at much higher temperatures and lower insolation. This makes it difficult to predict their performance under these varying conditions. In order to help PV system designers predict variations in the performance of PV modules with temperature, manufacturers provide a performance indicator called the nominal operating cell temperature (NOCT). The NOCT is defined as the temperature reached by the open-circuited cells in a module under the following conditions: insolation of 800(W/m^2), ambient temperature is 20°C, wind speed = 1 m/s, and open backside mounting. For any other ambient temperature, the cell temperature may be estimated using the expression [2]

$$T_{\text{Cell}} = T_{\text{ambient}} + \left(\frac{\text{NOCT} - 20\ ^{\circ}C}{800} \right) \cdot G \qquad (4.13)$$

Example 4.3

A PV module made from crystalline silicon has an open-circuit voltage of 21.5V and a maximum power of 120W at STC. The module has a NOCT of 45°C. Estimate the cell temperature, the open-circuit voltage, and the maximum power output for this module under insolation G = 1000 W/m^2 and ambient temperature of 30°C.

Note that for crystalline silicon, the open-circuit voltage drops by 0.37%/°C and the power drops by 0.5%/°C.

Solution

First, we need to find the temperature of the cell when the ambient temperature is 30°C from

$$T_{Cell} = T_{ambient} + \left(\frac{NOCT - 20\ ^{\circ}C}{800}\right) \cdot G$$

$$\therefore T_{Cell} = 30 + \left(\frac{45 - 20\ ^{\circ}C}{800}\right) \cdot 1000 = 61.25\ ^{\circ}C$$

The new open-circuit voltage is calculated from

$$V_{OCT2} = V_{OCT1}[1 - \alpha_V(T_2 - T_1)] \text{ where } T_1 = 25\ ^{\circ}C$$
$$V_{OCT2} = 21.5\ V\left[\ 1 - 0.0037(61.25 - 25)\ ^{\circ}C\right] = 18.62\ V$$

The new maximum power output is obtained from

$$P_{max} = 120\ W\left[\ 1 - 0.005(61.25 - 25)\ ^{\circ}C\right] = 98.25\ W$$

That is, the open-circuit voltage drops from 21.5V at 25°C to 18.62V at the actual operating cell temperature of 61.25°C, while the power delivered by the module is reduced from 120W to 98.25W representing a loss of about 17%.

4.7 Effects of Partial Shading on the I-V and P-V Characteristics and Bypass Diodes

In practice, it is difficult to ensure that all PV modules in an array or all cells in a single module have the same level of insolation, a phenomenon referred to as partial shading (PS). This phenomenon can be caused by obstacles such as clouds, dirt, buildings, trees, bird droppings, and snow. PS has adverse effects on the performance of PV systems and its mitigation requires the use of bypass diodes that give rise to multiple peaks in the P-V curve and consequently complicates the design of the maximum power point tracker, which is used to track the global maximum power point. A simple method of visualizing and analyzing the effect of partial shading is to consider a PV module, which consists of n cells (connected in series) as shown in Figure 4.10 with the nth cell represented by its single-diode equivalent circuit [3]. Under normal conditions of uniform insolation, the output voltage is V (V) and the output current is I (A). If the

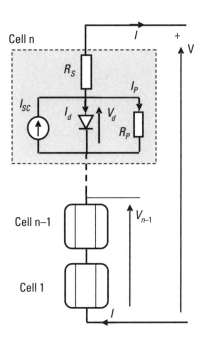

Figure 4.10   Effects of partial shading.

nth cell is completely shaded, it will not pass any current and its short-circuit current will be reduced to zero.

Consequently, all of the module current I (A) from the rest of the (n–1) cells will flow through the parallel and series resistors R_P and R_S of the shaded cell, causing a reverse voltage across the diode of the shaded cell. Hence, its diode will be reverse-biased, and the cell will not pass any current apart from the very small reverse-bias current of the diode. Under this shading condition, the terminal voltage of the module V_{Sh} will be the voltage due to the (n–1) remaining unshaded cells V_{n-1}, less the voltage drop due to the series and parallel resistors of the shaded cell, and hence, the module's voltage will be

$$V_{Sh} = V_{n-1} - I(R_S + R_P)$$

(4.14)

When there is no shading at all (i.e., all cells are receiving the same amount of sunlight) the terminal voltage of the module is V and the voltage of the lower (n–1) cells will be a fraction (n–1)/n of this voltage given as

$$V_{n-1} = \left(\frac{n-1}{n}\right)V$$

(4.15)

Therefore, under the condition that the nth cell is shaded the terminal voltage is

$$V_{Sh} = \left(\frac{n-1}{n}\right)V - I(R_S + R_P) \tag{4.16}$$

The loss (i.e., the drop in the voltage ΔV at any given current I caused by the shaded cell) is therefore given by

$$\begin{aligned}
\Delta V &= V - V_{Sh} \\
&= V - \left[\left(\frac{n-1}{n}\right)V - I(R_S + R_P)\right] \\
&= \frac{V}{n} + I(R_S + R_P)
\end{aligned} \tag{4.17}$$

However, in practice $R_P \gg R_S$, and hence we may use the approximation

$$\Delta V = \frac{V}{n} + IR_P \tag{4.18}$$

This equation gives the voltage drop at any given current I caused by the shading of one cell and its effect on the I-V characteristic of the module is illustrated in Figure 4.11, which indicates that the module voltage at any current I moves to the left by the amount ΔV; that is, the terminal voltage of the module is reduced from V to $V - \Delta V$.

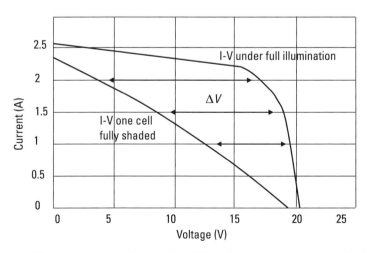

Figure 4.11   Effects of shading a single cell in a PV module of n series-connected cells.

Example 4.4

A PV module consists of 36 identical cells connected in series. Each cell has a parallel resistance of $R_p = 7\Omega$ and a series resistance $R_S = 5$ mΩ. In full sun the terminal voltage is $V = 20$V and the output current is $I = 2$A. If one cell is shaded and assuming that the output current remains the same, determine

    a. The new values of the output voltage and power of the module;

    b. The voltage drop across the shaded cell;

    c. The power dissipated in the shaded cell.

Solution:

$$\Delta V = \frac{V}{n} + IR_p = \frac{20}{36} + 2 \times 7 = 14.56 \ V$$

The new terminal voltage of the module will, therefore, be

$$V_{Sh} = V - \Delta V = 20 - 14.56 = 5.44 \ V$$

output power when in full sun is $VI = 20 \times 2 = 40 \, W$, and when one cell is shaded this power becomes 5.44V X 2A =10.88W . That is, the power is reduced by approximately 70%.

    b. The current of 2A will pass through the parallel and series resistances of the shaded cell; therefore the voltage drop across the shaded cell is

$$V_{Sh\_Cell} = I(R_S + R_p) = 2(0.005 + 7) = 14.01 \ V$$

When this cell is not shaded it will contribute to the module output voltage by 0.5V; however, when it is shaded it reduces the output voltage by 14.01V.

    c. The power dissipated in the shaded cell is $VI = 14.01 V X 2A = 28.01$W.

We conclude that, even with one cell out of 36 cells is shaded, the output power is reduced by about 70%. Further, the power dissipated in the shaded cell is converted to heat, which gives rise to a hot spot that may damage the cell and the module. The adverse effects of partial shading on the I-V characteristics are illustrated in Figure 4.12, which shows typical variations of the I-V curve when one and two cells are completely shaded.

Also included in Figure 4.12 is the load-line (dashed line) for charging a battery at 13.5V, which indicates that even with only one cell shaded, the charging current is reduced from about 2.5A to 0.8A. Therefore, measures must be taken to mitigate the effects of shading. This is normally done using anti-parallel bypass diodes as shown in Figure 4.13, which shows one cell in a PV module. Under normal conditions with no shading, the module current passes

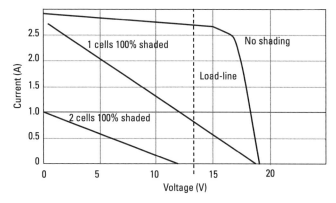

Figure 4.12 Effect of partial shading on the I-V characteristic of a PV module.

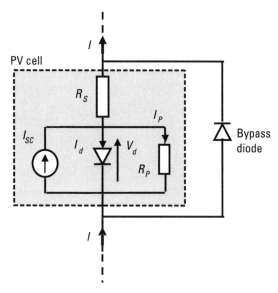

Figure 4.13 Bypass diodes in a PV module.

through the cell and the voltage rise across it will reverse-bias the bypass diode, and hence it acts just like an open circuit. However, when the cell is shaded, its light-generated current is zero and the module's current passes through its parallel and series resistances, which will give rise to a voltage drop across these resistances. This voltage will forward-bias the bypass diode, and hence, it will start conducting, thus passing the module current to the next cell bypassing the shaded one. Once forward-biased the voltage drop across the diode is only about 0.6V, and this is the voltage that the output voltage of the module will lose, which is a lot less than the 14V we have seen in the previous example.

Therefore, bypass diodes don't only alleviate the effects of shading on the voltage and current characteristics of a PV module, but also prevent hot spots. In practice, manufacturers don't manufacture modules with bypass diodes across each individual cell; instead, bypass diodes are usually placed across groups of series connected cells making up a string. That is, a PV module consists of number of strings equipped with one bypass diode for each string. There are two common configurations of bypass diode implemented on PV modules—overlapped and nonoverlapped schemes—as depicted in Figure 4.14 [4]. However, bypass diodes can lead to complex I-V and P-V characteristics; in particular it gives rise to multiple peaks in the P-V curve of a PV module/array, which complicates the design of MPPTs [5]. As shown in Figure 4.15, when there is no shading, there is only one global maximum power point. However, when one cell is shaded and its bypass diode is conducting, the P-V curve has an additional local MPP, which is due to the conducting bypass diode.

When connecting PV modules in series to make a PV array and when one module is shaded, similar effects occur as with the case of shading one cell in a module. That is, in an array of modules connected in series, a bypass diode is connected across each module to reduce the effects of shading. For example, as shown in Figure 4.16, when a module is shaded, the current is diverted into the antiparallel bypass diode.

One method of mitigating the adverse effects of partial shading on the performance of a PV array is to vary the configuration of the array depending on the existing shading pattern [6]. There are four common array configurations: series-parallel (SP), total-cross-tied (TCT), bridge-linked (BL), and honeycomb (HC), as shown in Figure 4.17 [7].

Details of the variations of the performance of a PV array with array configurations and the arrangements of bypass diodes within a module, can be found in [8, 9].

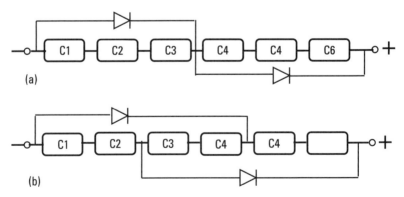

Figure 4.14   (a) Nonoverlapped and (b) overlapped configurations of bypass diodes.

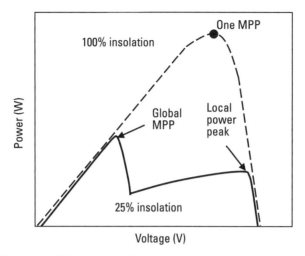

Figure 4.15 Effect of partial shading and bypass diode.

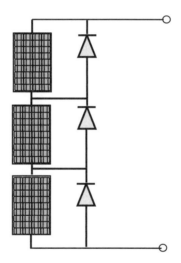

Figure 4.16 Using bypass diodes in a PV array.

4.8 Blocking Diode

To obtain larger current from a PV system we connect PV modules in parallel as shown in Figure 4.18. However, if one module is shaded or becomes malfunctioning, then current can flow in reverse through the malfunctioning module. Similarly, if the array is used to charge a battery, at night the battery will start supplying the PV array with current. To prevent reversal of current flow, blocking diodes are connected in series with each string as shown in Figure 4.18.

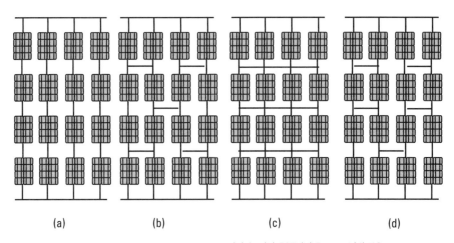

Figure 4.17   Different PV array configurations: (a) SP, (b) TCT, (c) BL, and (d) HC.

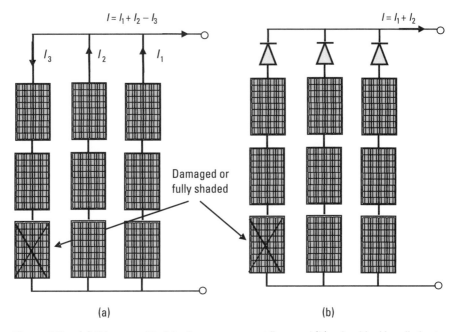

Figure 4.18   (a) PV array with risk of reverse current flow, and (b) using blocking diodes to eliminate this risk.

4.9   Summary

PV modules are always manufactured of several PV cells connected in series and are connected in series and/or parallel combinations to obtain a larger amount of power. The open-circuit voltage of a PV module is reduced considerably with increasing temperature but only slightly with increasing insolation. The short-circuit-current increases in direct proportion to insolation but increases only slightly with temperature. Partial shading can have adverse effects on the performance of a PV module/array and even when only one cell is shaded in a module, it can reduce the output power by more than 50%. Bypass diodes are used to mitigate the effects of° partial shading. Blocking diodes are used in series with PV modules to prevent current flowing back into faulty or shaded modules.

4.10 Problems

P4.1 a. Define the fill factor and explain its significance.

b. Explain with appropriate diagrams the variations of the I-V characteristic of a PV module with variations in insolation and temperature.

c. Describe using your own words and with the aid of appropriate diagrams the effects of partial shading on the electrical characteristics of a PV module.

d. Explain how the use of bypass diodes can reduce the adverse effects of partial shading.

e. Explain the reason for using blocking diodes in PV modules and arrays.

P4.2 Determine the cell temperature and power delivered by a 120W crystalline silicon   PV module under the following conditions. You may assume that the temperature coefficient for power loss is α_p = 0.005/°C.

a. Ambient temperature of 25°C, NOCT of 55°C and insolation of 1000 W/m².

b. Ambient temperature of 2°C, NOCT of 40°C and insolation of 400 W/m².

c. Ambient temperature of 35°C, NOCT of 50°C and insolation of 800 W/m².

P4.3 Determine the cell temperature and open-circuit voltage of a 22V photovoltaic module under the following conditions. You may assume that the temperature coefficient for voltage loss remains constant at α_V = 0.0037/°C.

a. Ambient temperature of 25°C and NOCT of 55°C, insolation of 1000 W/m².

b. Ambient temperature of 0°C and NOCT of 35°C, insolation of 400 W/m².

c. Ambient temperature of 40°C and NOCT of 55°C, insolation of 800 W/m².

P4.4   A PV module consists of 36 cells connected in series has a short-circuit current of 4.5A and an open-circuit voltage of 18V at STC. Assume that each cell has a parallel resistance of 4 ohms and negligible series resistance. Making any reasonable assumptions (a) draw the I-V characteristic if one cell is completely shaded, and (b) if the module is used to charge a 12V battery, determine the current that would be delivered to the battery under this shading condition.

P4.5   A PV module consists of two series-connected strings of cells. Each string has the same number of series-connected cells. The module has the I-V curve shown in Figure 4.19 under STC. The module has two bypass diodes one for each string. If one string is completely shaded, draw the new I-V curve under this condition if the bypass diodes are ideal (i.e., they have no voltage drop across them when conducting).

P4.6   A PV cell has the I-V characteristic shown in Figure 4.20 at STC. Sketch the characteristic of the combination of 4 of these cells when connected (a) in series and (b) in parallel.

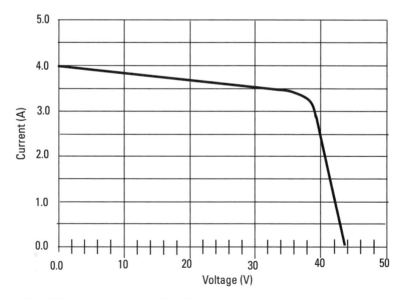

Figure 4.19   I-V curve for a generic PV module.

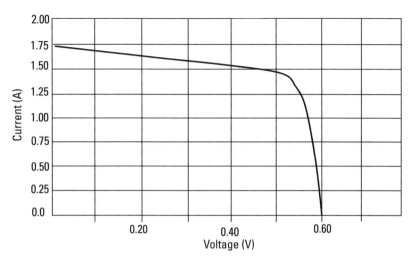

Figure 4.20   I-V curve for a generic PV cell.

P4.7   A PV cell has the I-V characteristic shown in Figure 4.20 at STC. Estimate the parallel and series resistance of the cell. Four of these cells are connected in series and a bypass diode is used across each cell. Making any reasonable assumptions, draw the I-V curve for the combination if only one cell is completely shaded.

P4.8   A PV cell has the I-V characteristic shown in Figure 4.20 at STC. It has an active area of 62 cm^2. Making any reasonable assumptions determine the fill factor and efficiency of the cell.

References

[1]   Mitsubishi, "Mitsubishi Electric Photovoltaic Module: PV-MF110EC3 PV module," 2004, https://www.mitsubishielectricsolar.com/images/uploads/documents/specs/MF_spec_sheet_110EC3_final.pdf.

[2]   Ross, R. G., "Flat-Plate Photovoltaic Array Design Optimization," *14th IEEE Photovoltaic Specialists Conference*, San Diego, CA, 1980.

[3]   Masters, G. M., *Renewable and Efficient Electric Power Systems*, New York: John Wiley & Sons, 2004.

[4]   Díaz-Dorado, E., A. Suarez-Garcia, C. Carrillo, and J. Cidra, "Influence of the Shadows in Photovoltaic Systems with Different Configurations of Bypass Diodes," *Int. Symp. on Power Electronics, Electrical drives, Automation and Motion*, Pisa, 2010.

[5]   Anani, N., M. Shahid, M. Al-kharji, and J. Ponciano, "A CAD Package for Modeling and Simulation of PV Arrays under Partial Shading Conditions," *Energy Procedia*, Vol. 42, 2013, pp. 397–405.

[6] Picault, D., B. Raison, S. Bacha, and A. Aguilera, "Forecasting Photovoltaic Array Power Production Subject to Mismatch Losses," *Solar Energy*, Vol. 84, No. 7, 2010, pp. 1301–1309.

[7] Humada, A. M., M. Herwan Bin Sulaiman, M. Hojabri, H. M. Hamada, and M. N. Ahmed, "A Review on Photovoltaic Array Behaviour, Configuration Strategies and Models under Mismatch Conditions," *ARPN Journal of Engineering and Applied Sciences*, Vol. 11, No. 7, 2016, pp. 4896–4903.

[8] Ibrahim, H., and N. Anani, "Performance of Different PV Array Configurations under Different Partial Shading Conditions," *International Conference on Sustainability in Energy and Buildings SEB-19*, Budapest, 2019.

[9] Ibrahim, H., and N. Anani, "Study of the Effect of Different Configurations of Bypass Diodes on the Performance of a PV String," *International Conference on Sustainability in Energy and Buildings SEB-19*, Budapest, 2019.

5

Photovoltaics: PV Generating Systems

5.1 Learning Outcomes

After actively engaging with the material in this chapter, you should be able to

1. Describe using block diagrams, the structure of stand-alone and grid-connected PV systems;

2. Explain the need for direct current-alternating current (DC-AC) and DC-DC converters in a PV system;

3. Draw the load-line for resistive load and determine the operating point on the I-V curve of a PV module;

4. Draw the load-line for battery load and determine the operating point on the I-V curve of a PV module;

5. Use the I-V curves to explain the operation of a DC motor load;

6. Discuss the need for a maximum power point tracking in a PV system;

7. Distinguish between deep and shallow discharge batteries;

8. Explain the need for blocking diodes in PV modules and arrays;

9. Calculate energy loss due to blocking diodes.

5.2   Overview

In this chapter, we will study different configurations of stand-alone and grid-connected PV generating systems and briefly look at the structure and operation of the DC-AC and DC-DC converters, which are common components in PV systems. We will also consider the issue of load matching and the use of maximum power point tracking to maximize the harvested energy from a given PV generator. The characteristics and operation of a PV generator with resistive, DC motor, and battery loads are explained. Finally, several problems are included at the end of chapter to consolidate understanding of the material.

5.3   Introduction

In general, a photovoltaic system consists of a PV generator, basically a module or an array, and a collection of electronic subsystems that are necessary to maximize the harvested energy and convert it into a usable form. There are two main configurations of PV systems: stand-alone and grid-connected.

5.4   Stand-alone Photovoltaic Generating Systems

A stand-alone, or off-grid, PV system is used mainly in remote areas where grid utility is not available or is not economically viable. In other words, a stand-alone system is an alternate to the grid. Off-grid systems are designed and sized to meet requirements of certain loads and applications, such as water pumping and ventilation systems, and if energy storage is available they can also be used for lighting during nighttime. There are a few different configurations of stand-alone PV systems as explained below.

5.4.1   Directly Coupled PV system

In a directly coupled PV system, a DC load is connected either directly or through a suitable interface to the PV generator, as indicated in Figure 5.1. This type of systems is suitable for application, such as water pumping and ventilation in remote areas.

Figure 5.1   Stand-alone PV system with a directly connected DC load.

Depending on the nature of the load and application, the controller could be as simple as a diode to prevent current flowing back into the PV generator or a MPPT to maximize the efficiency of the system by matching the load to the PV generator (i.e., the array/module).

5.4.2 Directly Coupled PV System with Battery

Alternatively, the generated electricity from the PV system may be used, in addition to supplying a DC load, to charge a bank of batteries. The DC voltage from the batteries can then be converted to AC voltage using a standard DC-AC inverter, as shown in Figure 5.2. The controller prevents reversal of the current when the battery is fully charged and normally includes a MPPT. This is clearly more efficient than a directly coupled system as it provides for both AC and DC loads. In addition, when the generated electricity is not needed it can be stored in the batteries for use during nighttime or when the weather is not conducive for generating energy. Note that in this system the DC-AC inverter must be fed from the battery and not from the PV generator.

5.4.3 Hybrid System

In a hybrid system, a conventional energy source, such as a diesel generator, is included in the system, as shown in Figure 5.3. This improves the security of the supply and results in a more reliable power plant.

The diesel generator represents a backup source of energy so that when batteries are depleted and weather is not allowing for enough energy generation, the generator can be used as an alternative supply. In addition, it may be used to support peak load demands. The AC-DC converter converts the AC voltage from the diesel generator into DC to charge the batteries when needed. A hybrid system may also include other forms of renewable energy generators, such as a wind turbine.

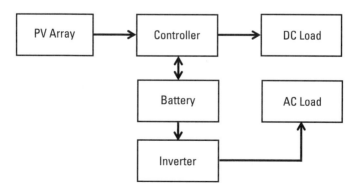

Figure 5.2 Stand-alone PV system with a DC and AC supply.

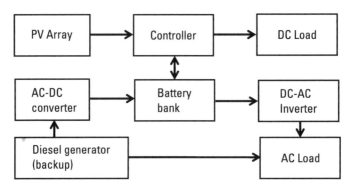

Figure 5.3 Hybrid stand-alone PV system.

5.5 Grid-Connected Systems

Stand-alone systems are suitable for remote areas where a grid is not available or is economically not viable. However, when a grid is available, the utilization of a PV system can be enhanced by integrating it with the grid. In a grid-connected system, such as depicted in Figure 5.4, the load can be met by the grid when weather conditions are not conducive for generating electricity, and if batteries are not fully charged. On the other hand, when the harvested renewable energy is in excess of load demand, the excess energy can be fed back or sold to the grid. The DC-AC grid-tie inverter is used to convert a DC voltage into an AC voltage that is compatible with the grid voltage. In addition, the grid-tie inverter has some additional functions compared to a standard inverter. For example, it is equipped with a phase-locked loop (PLL) that allows it to synchronize its output voltage waveform with the grid voltage to ensure that when the inverter's output voltage is connected to the grid, it has the same frequency and phase [1]. Furthermore, a grid inverter is required by law to have certain features and

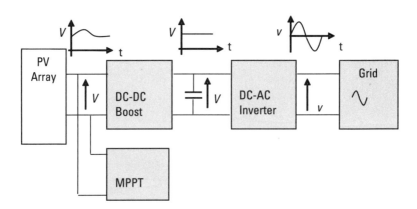

Figure 5.4 Grid-connected PV system.

capabilities for safety and power quality. For example, in the event of a power outage, the inverter must be capable of detecting power loss of the grid and automatically switching itself off.

5.5.1   DC-AC Inverters

The concept of inversion (i.e., converting a DC voltage into an AC voltage) is very simple. Consider the circuit arrangement shown in Figure 5.5, which is called a full-bridge single-phase DC-AC inverter. The switches S1-S4 are, in practice, semiconductor devices like transistors, which are used as controllable electronic switches. The circuit can be used to generate square or quasi-square output voltage waveforms. When switches S1 and S2 are turned on for half-cycle, $T_o/2$, while S3 and S4 are turned off, point A is connected to the positive rail, while point B is connected to the negative rail of the of DC supply voltage. Neglecting the on-state voltage drop across the switches, the load voltage is the same as the supply (i.e., $V_L = V_{dc}$). Similarly, during the next half-cycle, switches S3 and S4 are turned on while S1 and S2 are switched off, point A is connected to the negative rail, and point B is connected to the positive rail of the DC supply; hence, the load voltage is $V_L = -V_{dc}$. If this process is repeated periodically, and if the switching action occupies a negligible fraction of the half-cycle, the load voltage is a periodic square waveform, as shown in Figure 5.6, whose amplitude is $\pm V_{dc}$.

The frequency of the square wave (i.e., the number of cycles per second) is

$$f_o = \frac{1}{T_o} \tag{5.1}$$

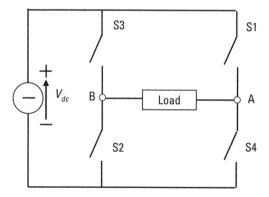

Figure 5.5   Single-phase full-bridge inverter.

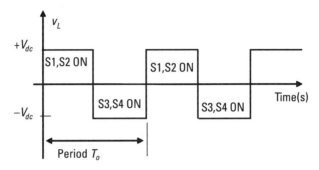

Figure 5.6 Output voltage waveform of the single-phase inverter.

For example, if S1 and S2 are turned on for 10ms (while S3 and S4 are turned off) and S3 and S4 are turned on for the 10ms (while S1 and S2 are turned off), the frequency of the square wave will be 50 Hz (1/0.02). This is called the fundamental frequency and the period $T_o = 20$ ms is the fundamental period.

Further analysis can be used to show that this square wave consists of an infinite sum of sinusoidal voltage components called the spectrum of the square wave signal, which are given by [2]

$$v_L = \frac{4V_{dc}}{(2n-1)\pi} \sum_{n=1}^{\infty} \sin(2n-1)\omega_o t \qquad (5.2)$$

The angular frequency ω, in radians per second, is defined as

$$\omega_o = 2\pi f_o \qquad (5.3)$$

Expanding the above expression for the load voltage, we can write

$$v_L = \frac{4V_{dc}}{\pi} \sin \omega_o t + \frac{4V_{dc}}{3\pi} \sin 3\omega_o t + \frac{4V_{dc}}{5\pi} \sin 5\omega_o t + \frac{4V_{dc}}{7\pi} \sin 7\omega_o t + \ldots\ldots$$

The first voltage v_1 is called the fundamental voltage, which is shown in Figure 5.7, and has the same frequency as the original square wave

$$v_1 = \frac{4V_{dc}}{\pi} \sin \omega_o t = V_{1m} \sin 2\pi f_o t \qquad (5.4)$$

The maximum or peak value of the fundamental component is

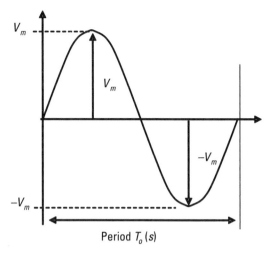

Figure 5.7 Fundamental sinewave of the square wave.

$$V_{1m} = \frac{4V_{dc}}{\pi} = 1.27 \, V_{dc} \tag{5.5}$$

The rest of the component voltages are at multiple frequencies of the fundamental frequency and these are called harmonic voltages. For a square wave, only odd harmonics exist (i.e., the third, fifth, seventh, and so on). In a real inverter, these harmonics are eliminated so that the output is only the fundamental voltage.

Example 5.1

A single-phase full-bridge inverter is fed from a 100V DC source. The fundamental switching period is 20 ms. Write down sinusoidal expressions for the fundamental, the third, and the fifth harmonic voltages.

Solution:

$T_0 = 20$ ms. Therefore, the fundaments switching frequency is

$$f_o = \frac{1}{0.02} = 50 \text{ Hz}$$

The fundamental angular frequency is

$$\omega_o = 2\pi f_o = 100\pi \text{ rad/s}$$

The amplitude of the fundamental voltage is

$$V_{1m} = \frac{4V_{dc}}{\pi} = 1.27(100) = 127 \text{ V}$$

So $v_1 = 127 \sin(100\pi t)$.
The amplitude of the third harmonic is

$$V_{3m} = \frac{4V_{dc}}{3\pi} = 42.4 \text{ V}$$

Hence, $v_3 = 42.4 \sin(300\pi t)$.
The amplitude of the fifth harmonic is

$$V_{5m} = \frac{4V_{dc}}{5\pi} = 25.5 \text{ V}$$

Therefore, $v_5 = 25.46 \sin(500\pi t)$.

5.5.2   Standards for Grid-Inverters

There are several factors that influence the operation of a grid-tie inverter, such as the response time, power factor, harmonics and DC injection into the grid, synchronization, and protection. In the United Kingdom, the G83 standard sets the regulations that a grid-tie inverter must satisfy before it can be interfaced to the grid [3]. A grid-tie inverter must provide clean sinusoidal voltage (i.e., free from harmonics) and this voltage must be synchronized with the grid voltage (i.e., the inverter's voltage must track the frequency, amplitude, and phase of the grid voltage). In the United States, the IEEE 1547 standard covers and establishes the specifications and requirements for the interconnection and interoperability between any distributed energy resources and the electric power systems [4]. It provides requirements relevant to the performance (e.g., voltage regulation, operation, testing, safety considerations, and maintenance of the interconnection). For example, in the event of a fault on the grid, the inverter must not be allowed to continue back-feeding part (an island) of the grid as this can be dangerous to engineers and technician who might be working on the presumably dead part of the grid. Therefore, a PV inverter must be capable of automatically switching itself off in the event of a power-down, whether intentional or due to a fault. However, in some systems, intentional islanding is allowed when steps are taken to divert power away from the faulty location of the grid. Unintentional islanding refers to the grid-connected inverter continuing to supply power to the grid during the event of a fault on the grid that is not allowed. In addition, a grid-connected inverter must be capable of detecting abnormal conditions, such as a sudden change in system frequency and/or voltage or a sudden increase in reactive or active power, and disconnects itself from the grid under such conditions [5].

5.5.3 DC-DC Converters

Since the output from a PV array or module is typically low-voltage (e.g., 12V, 24V, or 48V) compared to the grid voltage, a DC-DC boost converter is normally used before the DC-AC inverter stage to step up the PV voltage. The basic building block of a DC-DC boost, also known as step-up, converter is shown in Figure 5.8. The switch is a semiconductor device, such as a bipolar junction transistor (BJT), which can be operated as a controllable electronic switch using a switching voltage signal, as shown in Figure 5.9. For a BJT switch, the switching signal typically alternates between 0 and 5V.

When the signal is 5V, the BJT switch is turned on and when it is 0V, the BJT switch is turned off. We define the duty cycle (*D*) of the switching waveform as the ratio of the on-time to the period

$$D = \frac{t_{\text{on}}}{t_{\text{on}} + t_{\text{off}}} \tag{5.6}$$

Now suppose that the converter has been running for some time and there is an output voltage, which is the same as the capacitor voltage, and the

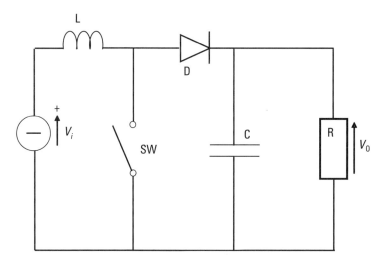

Figure 5.8 DC-DC boost converter.

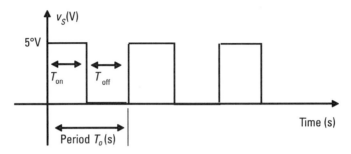

Figure 5.9   Switching waveform.

inductor has some energy stored in its magnetic field. The diode will be reverse-biased, thus preventing current flow from the load back into the inductor. When the switch is closed, the input voltage appears across the inductor and current flows through the inductor down the switch and back to the DC source. During this time, t_{on}, the inductor stores energy in its magnetic field. When the switch is opened, the inductor induces a relatively high back EMF, which will try to maintain the current at its previous level and thus pushes a spike of current through the diode, which will increase the capacitor voltage; hence, the output voltage is increased. The output voltage is given in terms of the duty cycle and the input voltage as [6]

$$V_o = \frac{V_i}{1-D} \tag{5.7}$$

This equation indicates that the output voltage is always higher than the input voltage for practical values of duty cycle (D) greater than zero and less the unity. In practice, a controller generates the switching waveform shown above with varying duty cycle. The duty cycle is normally varied by varying the on-time (i.e., the width of the pulse) keeping the period constant. The switching waveform is called a pulse width modulation (PWM) signal, and the controller is a PWM generator. The duty cycle is used as a control parameter for the DC-DC converter to set its output voltage as required by the PWM controller. Another type of DC-DC converter is the step-down converter, which is used to step an input voltage down according to the relation [6]

$$V_o = DV_i \tag{5.8}$$

Since the duty cycle, D, can take any value between 0 and 1, in theory a step-down (also known as Buck) converter can provide an output voltage V_o from 0 to V_i.

5.6 Operation of a PV Generator with Different Types of Loads

The I-V curves for a PV module or array tell us the range of voltages and currents over which a PV module can operate. The actual operating point depends on the nature of the load. To find the operating point of a particular load, all we have to do is to plot the I-V curve for the load on the same axes as the I-V curve for the PV generator. The point of intersection of the two curves will be the operating point of the generator. We will consider three different types of loads: pure resistor, battery, and a DC motor.

5.6.1 Load Curve for a Purely Resistive Load

In Figure 5.10, a variable load resistor, *R*, is connected directly to a PV module. We need to determine its operating point. To do this we draw the load-line (i.e., the I-V equation for the load resistor) on the same axes as the I-V curve of the PV module.

The equation of the load-line for a resistor, *R*, is simply

$$I = \frac{1}{R}V \qquad (5.9)$$

The value of the resistance *R* is the reciprocal of the slope of the load-line drawn on the I-V axes, as shown in Figure 5.10, where three load-lines for three different resistance values are drawn. Maximum power from the PV generator is delivered when the resistance of the load is the same as the resistance looking into the PV generator (i.e., the resistance of the power source R_{PV}). This

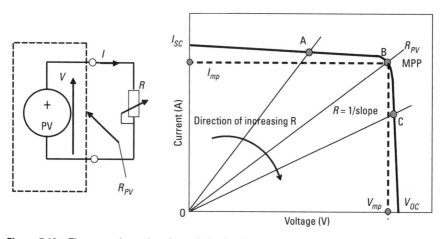

Figure 5.10 The operating point of a resistive load.

condition is shown by the load-line OB, which corresponds to the case $R = R_{PV}$ where the PV generator is operating at its maximum power point. In this case, we say that the load resistance is matched to the generator. If the load resistance is now decreased below the resistance of the generator R_{PV}, the load-line moves to the left and the operating moves to point A, for example, where the power is less than the MPP. Similarly, if the load resistance is increased above the value of the internal resistance of the PV generator, the load-line moves to the right and the operating point moves to point C, for example, where the power is again less than the MPP.

Therefore, for a PV generator to operate at its maximum power point, the generator must be connected to a load whose resistance is the same as the internal resistance of the generator. The process of matching the resistance of the load to the resistance of the generator is called resistance, or more generally, impedance matching. However, in practice a load is not fixed, so the solution for impedance matching is to introduce a power processing system between the load and the PV generator, as shown in Figure 5.11, whose function is to provide impedance matching. The power processing system has a control input that is used to change its input resistance, R_{in}, so that it is always equal to the resistance of the generator, R_{PV}, regardless of the value of the load. Typically, the power processing system is a DC-DC converter and the control input is its duty cycle. Note that the DC-DC converter is used here for a different purpose than the normal function of stepping its input voltage up or down. Furthermore, the maximum power point, and hence, the internal resistance of a PV generator,

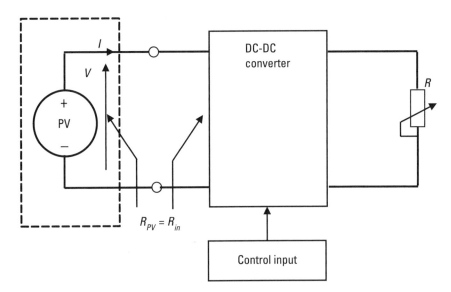

Figure 5.11   Using a DC-DC converter to provide resistance matching.

changes with insolation, temperature, and with aging. Variation of the MPP with insolation is illustrated in Figure 5.12.

This shows that the operating point slips off the MPP as insolation changes. Therefore, we need some mechanism by which we can detect the present MPP and accordingly set the control input to the DC-DC converter to set its input impedance at the value needed to draw maximum power from the PV generator. This is called MPPT, which we will consider soon.

5.6.2   Load Curve for a DC Motor

PV generators, arrays, or modules are in principle DC sources, which makes them suitable for driving DC motors. Such motors are used in many applications, such as water pumping, refrigeration, and ventilation. A DC motor runs at almost a constant speed for a given supply voltage and may be represented by the equivalent circuit shown in Figure 5.13, which also includes the I-V terminal characteristic of a typical permanent magnet DC motor. The basic equation for a permanent magnet DC motor in the steady state is

$$V = IR_a + E_a \tag{5.10}$$

In the above equation, V(V) and I(A) are the motor input voltage and current, respectively, R_a is the armature resistance, and E_a is the armature induced EMF, which is proportional to the angular speed ω (rad/s) of the motor (i.e., if k is the back EMF constant of the motor) then

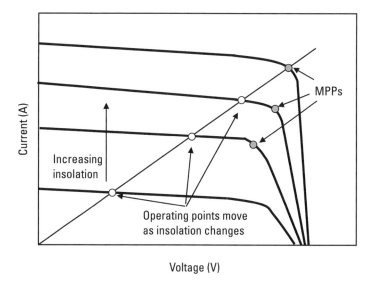

Figure 5.12   As the insolation changes, the operating point slips of the MPP for the same load.

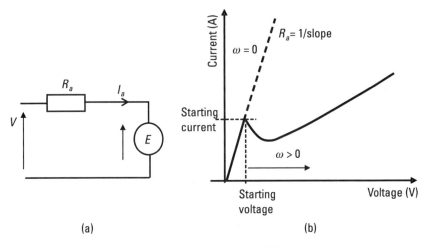

Figure 5.13   Equivalent circuit and I-V characteristic of a DC motor.

$$E_a = k\omega \tag{5.11}$$

Hence, we can rewrite the motor equation as

$$V = IR_a + k\omega \tag{5.12}$$

Multiplying both sides by the armature current

$$VI_a = I_a^2 R_a + kI_a\omega \tag{5.13}$$

This is the same as writing

$$P_{in} = P_{Loss} + P_m \tag{5.14}$$

The quantity VI_a represents the electrical input power, P_{in}, to the motor and $I_a^2 R_a$ is the power dissipation, P_{Loss}, in the armature resistance. The quantity $kI_a\omega$ is the mechanical output power, P_m, developed by the motor. The torque developed by the motor is

$$T = \frac{P_m}{\omega} \tag{5.15}$$

Hence, the motor torque is

$$T = \frac{kI_a\omega}{\omega} = kI_a \tag{5.16}$$

That is, the torque is proportional to the armature current and therefore when a motor is driven by a PV generator, there must be enough insolation to generate the minimum current required to start the motor, otherwise the motor won't start.

Once a DC motor starts, it runs at almost constant speed. When a load is added to the motor, the speed is reduced and consequently the back EMF is also reduced. This increases the current to increase the torque and the motor goes back to its original constant speed. Looking at the I-V curve in Figure 5.13, when the motor is switched on the current rises rapidly (high starting current low starting voltage) before the motor starts to turn. During this interval when the speed is 0 the gradient of the line is the reciprocal of the armature resistance. Once the motor starts turning the current rises but at a slower rate with increasing voltage. If we draw the motor I-V curve on the same axes as the I-V curves of the PV module, as shown in Figure 5.14, it is clear that the starting current of the motor requires approximately 350 W/m² of insolation. Once it is running, it only requires about 250 W/m². This means that the motor may not start even if the insolation is higher than that say 300 W/m². At lower insolation (e.g., in early morning) the operating point of the motor is far away from the knee of the I-V curve of the PV module where maximum power is available. However, although the available power may be sufficient to overcome the motor inertia and get it running, this power comes at low current and higher voltage; hence, it won't start the motor. However, if we use a linear current booster, which is an electronic circuit that delivers the same input power but at higher current and lower voltage, between the PV and the motor, the motor will be able to start running at lower insolation.

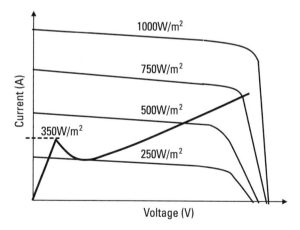

Figure 5.14 IV curve of a DC motor superimposed on the IV curve of a PV system.

Example 5.2

A permanent magnet DC motor has an armature resistance of 1 ohm. When it is spinning at its rated speed of 52.4 rad/s, the back EMF is 10V and the armature current is 2A.

Determine

a. The motor applied voltage;
b. The motor back EMF constant;
c. The electrical input and mechanical output powers;
d. The steady-state torque.

Solution:

$$V = E + I_a R_a = 10 + 2 \times 1 = 12 \text{ V}$$

Using $E_a = k\omega$, we obtain k as

$$k = \frac{E_a}{\omega} = \frac{10}{52.4} = 0.191 \text{ V.s/rad}$$

The electrical input power is

$$VI_a = 12 \text{ V} \times 2 \text{ A} = 24 \text{ W}$$

The mechanical power output is

$$EI_a = 10 \text{ V} \times 2 \text{ A} = 20 \text{ W}$$

(d) the steady-state torque,

$$T = kI_a = 0.191 \text{ (V.s/rad)} \times 2 \text{ A} = 0.382 \text{ Nm}$$

5.6.3   Load Curve for a Battery

Batteries are the most common devices used for storage of electrical energy. They are used to support peak demands and provide constant voltage. The I-V curve for an ideal battery is a vertical line whose equation is $V = E$, where E is the open-circuit voltage, or the EMF, of the battery, as shown in Figure 5.15. However, a real battery has, although small, internal resistance; hence, it is represented by an ideal battery in series with a resistance $R_B(\Omega)$ representing

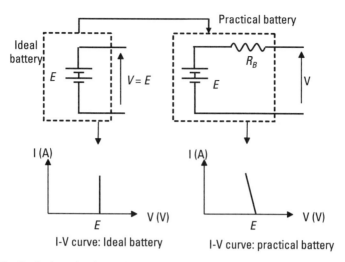

Figure 5.15 Equivalent circuit and IV characteristic for an ideal and practical battery.

its equivalent resistance. When a load is connected across a battery drawing a current I, the terminal voltage of a practical battery is therefore given as

$$V = E - IR_B \tag{5.17}$$

or

$$I = \frac{E}{R_B} - \frac{1}{R_B}V \tag{5.18}$$

This is the equation of the I-V curve for a real battery, which is also shown in Figure 5.15.

If we let the PV voltage be $V\,(V)$, then during the charging process the current is into the battery (Figure 5.16) and

$$V = IR_B + V_B \tag{5.19}$$

During the discharge process the I-V line of the battery is tilted to the left and

$$V_B = V + IR_B \tag{5.20}$$

However, for this case, the PV system is equipped with a charge controller that prevents the discharge of current from the battery into the PV modules when their voltage V is less than that of the battery. During the charging

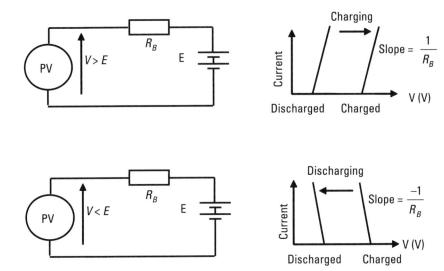

Figure 5.16 The process of charging and discharging.

process, the I-V curve of a battery moves to the right as the battery is charged by virtue of the insolation (i.e., during the daytime).

This means that there is a chance that the operating point may fall off the knee, particularly when the level of insolation is reduced since the knee moves to the left, as shown in Figure 5.17. This may not be an undesirable thing since current must be slowed or even cut off completely when a battery reaches full charge. In practice, the charge controller will automatically prevent overcharging of the batteries.

Example 5.3

A depleted 12V lead-acid battery has an open-circuit voltage of 11.2V and an internal resistance of 0.05 ohms. (a) The battery is now coupled to a PV module and receiving a current of 6A. Determine the module operating voltage. (b) The battery is now fully charged and has an open-circuit voltage of 12.8V and is still connected to the PV module. If the battery is delivering a current of 10A into a load, determine the module operating voltage.

Solution:

a. The PV terminal voltage of the module is

$$V = IR_B + V_B = (6A)(0.05\ \Omega) + 11.2 = 11.5\ V$$

b. The load (i.e., the terminal voltage) of the battery is

$$V_L = V_{OC} - IR_B = 12.8 - 10(0.05) = 12.3\ V$$

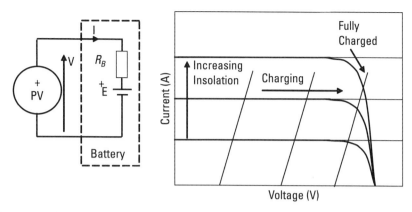

Figure 5.17 The charging process.

The operating voltage of the PV module is the same as the voltage across the battery (i.e., 12.3V).

5.7 Maximum Power Point Trackers

We have seen that certain loads, such as a purely resistive load, may be coupled directly to a PV generator while others, such as a DC-AC inverter may not. Instead, an inverter must be coupled to a PV generator via a special power processing system that includes a boost DC-DC converter. We have also seen that in order to operate a PV generator at its maximum efficiency, it must be operated at its MPP. However, as we have learned, the I-V characteristic of a PV generator depends on the uncontrollable climatic conditions, namely temperature and insolation, and the variation in these conditions makes the position of the maximum power point varying over a wide region of the I-V plane. As a consequence, the concept of connecting a power processing system or a load directly to a PV generator, forcing a constant voltage at its terminals sounds simple enough, but from an energy point of view can be very inefficient. For example, direct connection of a battery to a PV module would force the PV generator to operate at a constant voltage that may not be anywhere near the MPP. Similarly, if a PV generator is directly connected to a load resistance, as we have seen, there is no guarantee that the operating can be close to the MPP. Therefore, an interfacing stage between the PV generator and the load, or the power processing system, that uses or processes the power delivered by the PV generator is required. This stage is called the MPPT. The idea is that the MPPT must be capable of dynamically adapting the values of its input voltage and current to the instantaneous values of the PV generator at the prevailing MPP, while simultaneously maintaining its output current and voltage at the levels required by the load. This adaptation must be performed dynamically; that is, the MPPT must be capable of self-adjusting its parameters at run-time so as to change its input voltage and or current according to the current position of the MPP. A block diagram of a MPPT system is shown in Figure 5.18. It consists of a DC-DC converter, voltage and current sensors, and a MPPT controller. The controller consists of a microcontroller that has been programmed with a MPPT algorithm, which detects the current MPP. Once the MPPT algorithm detects the current MPP, it determines a suitable value for the control signal (i.e., the duty cycle of the converter). The microcontroller then generates and sends this value to the DC-DC converter so that its input resistance is adjusted to match the current optimum resistance of the PV generator, thus forcing the generator's operating point at the MPP.

Several algorithms have been published in the literature for detecting the current MPPT, which are based on various techniques, such as neural networks [7], fuzzy logic [8], biological swarm chasing algorithm [9], perturb and observe (P&O), and incremental conductance algorithms [10].

5.8   Batteries and Energy Storage

Stand-alone systems clearly require a system to store any harvested energy. Various technologies are available, such as compressed air, flywheels, hydrogen production, potential energy, and of course batteries [11]. In fact, the battery remains the most common method of electrical energy storage. There are many technologies of batteries, such as nickel-cadmium, nickel-metal hydride, lithium-ion, lithium-polymer, and nickel-zinc [12]. However, the simple lead-acid battery remains the workhorse of all batteries. There are other advantages

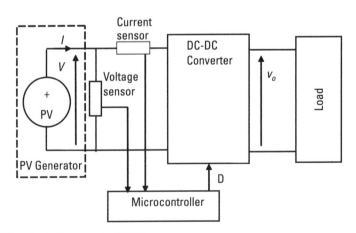

Figure 5.18   Block diagram of a MPPT system.

of using batteries in a PV system other than storage. These advantages include providing surge currents that are much higher than the instantaneous current available from a PV generator, as well as the intrinsic and automatic property of the battery in controlling the output voltage of the array so that loads voltages that are maintained within their own acceptable range [13]. A typical 12V lead-acid battery consists of six individual cells, each of 2V. A lead-acid battery is considered discharged once the voltage of each individual cell is down to 1.75V (i.e., the total battery voltage is about 10.5V ($6 \times 1.75 = 10.5V$)). The energy storage capacity of a battery is measured in the units of ampere-hour (Ah) at a nominal voltage and a given discharge rate. For example, the specifications for a lead-acid battery include the ampere-hour capacity (Ah) at a discharge rate that would drain the battery down to 10.5V over a specified length of time at an ambient temperature of 25°C. For example, a 12V lead-acid battery may be specified to have 20h, 400 Ah capacity. This means that this battery can deliver 20A (400 Ah/20 h) for 20 hours, and by the end of the 20 hours, the battery voltage will be reduced to 10.5V. In battery terminology, the C rating is used. When a 400-Ah battery is delivering 20A, it is said to be discharging at a C/20 rate. That is, the C refers to the Ah capacity and the 20 refers to the hours it will take to discharge the battery (400/20). Therefore, if the above 400-Ah battery is discharged at a rate of C/100, then the discharge current would 4A.

Lead-acid batteries are usually employed in remote areas with stand-alone PV systems mainly because they are relatively cheap. In deciding what type of batteries to use, a number of factors must be considered that include cost, lifetime, self-discharge, deep discharge, charging, and discharge current. Normally, battery manufacturers specify the mean value of the number of charge and discharge cycles of a battery as a function of the depth of discharge (DoD). The charge discharge cycle, or simply a battery cycle, involves completely draining and then fully charging the battery. The depth of discharge of a battery refers to the percentage of the battery that has been discharged relative to its full capacity, so a DoD of 100% means that the battery is empty, while a DoD of 0% indicates the battery is fully charged. An alternative method for measuring the capacity of a battery is the state-of-charge (SoC) which is opposite to the DoD. A 100% SoC means the battery is fully charged and 0% SoC means the battery is completely depleted. The number of battery cycles of a battery deteriorates rapidly with the DoD. Hence, in general the DoD should be kept as low as possible [14]. According to the DoD batteries are classified as either deep or shallow discharge. A deep discharge battery is designed to be regularly deeply discharged like traction batteries. A shallow discharge battery on the other hand, is designed to provide short, high current bursts like conventional automotive batteries. If a shallow discharge battery is constantly deeply discharged, its lifespan will be severely shortened. For this reason, deep discharge batteries are better suited for photovoltaic systems.

Example 5.4

A 200-Ah, 12V battery with an open-circuit voltage of 12.4V at its current SoC is charged at a C/5 rate with a charging voltage of 13.5V. (a) Determine the internal resistance of the battery. (b) What percentage of the battery input power is dissipated as heat in the internal resistance of the battery? (c) If the battery is charged at a C/20 rate, what percentage of the battery input power is dissipated as heat in the internal resistance of the battery?

Solution:

a. At C/5 the charging current is $\dfrac{200\ Ah}{5} = 40\ A$. Hence, the internal resistance is

$$R_{B} = \frac{V_{in} - V_{OC}}{40\ A} = \frac{13.5 - 12.4}{40\ A} = 0.0275\ \Omega$$

b. Therefore, the percentage power loss in the internal resistance of the battery is

$$\frac{I^2 R_B}{V_{in} I} \times 100\% = \frac{IR_B}{V_{in}} x100\% = \frac{40A \times 0.0275}{13.5} \times 100\% = 8.15\%$$

c. At C/20 the charging current is

$$\frac{200\ Ah}{20} = 10\ A$$

The input voltage is

$$V_{in} = V_{OC} + IR_B = 12.4 + (10A)(0.0275\ \Omega) = 12.675\ \text{V}$$

The percentage power losses are

$$\frac{10\ A \times 0.0275\ \Omega}{12.675\ V} \times 100\% = 2.17\%$$

Hence, charging at a lower current reduces the power loss in the internal resistance of the battery.

5.9 Effect of Blocking Diodes

A basic PV system that can be used at nighttime as well as during daytime consists of a PV module, a battery, and a load, as shown in Figure 5.19(a). This simple system will work satisfactorily and will operate something like a lightbulb at nighttime as long as the battery is not allowed to be overcharge or discharge deeply. One problem with the system is that during nighttime, the battery will discharge through the PV module. This problem may be alleviated using a blocking diode, as shown in shown in Figure 5.19(b). However, during daytime when the PV module is charging the battery there will be power loss in the diode.

During nighttime there is no short-circuit current, and ignoring the series resistance, the equivalent circuit of one cell of the n-cell module is shown in Figure 5.20.

And at nighttime without the blocking diode, the discharge current through the PV module is

$$I_B = I_d + I_P \tag{5.21}$$

Without any blocking diodes, the current through the cell diode at 25°C is

$$I_d = I_o(e^{38.9\,V_d} - 1) \tag{5.22}$$

The current through the parallel resistor is

$$I_P = \frac{V_d}{R_P} \tag{5.23}$$

That is

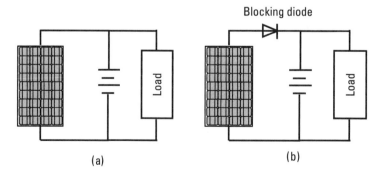

Blocking diode

(a) (b)

Figure 5.19 (a) A PV battery system feeding DC load, and (b) the same system, but with a blocking diode between the PV module and the battery.

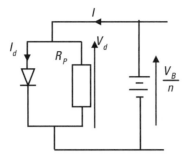

Figure 5.20 Equivalent circuit of one cell during nighttime.

$$I_B = I_o(e^{38.9\,V_\mathrm{d}} - 1) + \frac{V_\mathrm{d}}{R_\mathrm{P}} \qquad (5.24)$$

Let us now consider a practical example.

Example 5.5

A PV module is made up of 36 identical cells, each having a parallel resistance of 6 ohms and a reverse saturation current of $I_o = 5 \times 10^{-10}$A. On average, the module provides 4A for 7 hours each day. The module is connected without a blocking diode to a battery with voltage 12.7V.

a. Estimate the ampere-hours that will be discharged from the battery over a 12-hour night and the energy in watt-hours that will be lost due to this discharge.

b. If a blocking diode is added, how much energy will be dissipated through the diode during the daytime assuming the on-state voltage of the diode is 0.6V?

Solution:
The voltage across each cell is

$$V_\mathrm{d} = \frac{12.7}{36} = 0.353 \text{ V}$$

a. During nighttime, the current discharged from the battery is

$$I_B = 5 \times 10^{-10}(e^{38.9 \times 0.353} - 1) + \frac{0.353 \text{ V}}{6\,\Omega} = 59 \text{ mA}$$

So nighttime loss in Ah is

$$59 \text{ mA} \times 12 \text{ h} = 0.711 \text{ Ah}$$

At the battery voltage of 12.7V, the energy loss at night is

$$\text{Night time loss} = 0.71 \text{ Ah} \times 12.7 \text{ V} = 9.03 \text{ Wh}$$

b. During daytime the PV module should deliver

$$7 \text{ h} \times 4 \text{ } A = 28 \text{ Ah}$$

The nighttime loss w/out blocking diode is

$$0.711 \text{ Ah} / 28 \text{ Ah} = 0.021, \text{ i.e. } 2.1\%$$

With the blocking diode, the daytime loss due to the diode is

$$28 \text{ Ah} \times 0.6 \text{ V} = 16.8 \text{ Wh}$$

From the above example, the blocking diode loses more energy during the day while it is conducting than it saves during nighttime. In terms of the ampere-hour, without a blocking diode, only about 3% of the daytime solar energy gain is wasted during nighttime. However, in practice, the battery load-line will be some distance to the left away from the knee of the I-V curve of the PV module. Adding a blocking diode will move the load-line by a diode forward voltage drop (i.e., by about 0.6V) to the right in the direction of the knee, which is beneficial and will not cause any drop in the charging current. Therefore, despite the nighttime loss, adding a blocking diode is generally beneficial.

5.10 Summary

Photovoltaic generation systems can be stand-alone, hybrid, or grid-connected. Stand-alone systems are used in remote areas where the grid is not available while grid-connected systems offer better utilization of the harvested energy. A grid-connected inverter must satisfy certain regulations before it can be connected to the utility grid for reasons to do with safety and power quality. A PV generator may be connected directly to a load; however, this can be very inefficient and in practice a maximum power point tracker must be used to force a PV generator to operate at its maximum efficiency. Batteries remain the most attractive and easily available means of electrical energy storage. They improve the utilization of the harvested energy, but they are costly and require regular maintenance and inspection.

5.11   Problems

P5.1   a. Draw a block diagram of grid-connected PV system and explain the function of each block.

b. Sketch the variations of the I-V curve for a generic PV module with insolation;

c. Explain why a power processing system is normally needed as an interface in a PV generating system;

d. Explain the function of a MPPT in a PV system.

P5.2   A resistive load of 100 ohms is fed from a single-phase full-bridge DC-AC inverter, which is fed from a 200V DC source. Sketch the output voltage waveform and calculate the amplitudes of the fundamental and the third harmonic.

P5.3   A nearly depleted 12V lead-acid battery has an open-circuit voltage of 11.5V and an internal resistance of 0.06 ohms. (a) The battery is now coupled to a PV module and receiving a current of 5A. Determine the module operating voltage. (b) The battery is now fully charged and has an open-circuit voltage of 12.7V, and it is still connected to the PV module. If the battery is delivering a current of 10A into a load, determine the module operating voltage.

P5.4   A 300-Ah, 12V battery with an open-circuit voltage of 12.5V at its current SoC is charged at a C/10 rate with a charging voltage of 13.5V. (a) Determine the internal resistance of the battery and (b) what percentage of the battery input power is dissipated as heat in the internal resistance of the battery. (c) If the battery is charged at a C/20 rate, what percentage of the battery input power is dissipated as heat in the internal resistance of the battery?

P5.5   A PV module is made up of 36 identical cells, each having a parallel resistance of 8 ohms and a reverse saturation current of 100 pA. On average, the module provides 5A for 8 hours each day. The module is connected without a blocking diode to a battery with voltage 12.6V.

a. Estimate the ampere-hours that will be discharged from the battery over a 14-hour night and the energy in watt-hours that will be lost due to this discharge.

b. If a blocking diode is added, how much energy will be dissipated through the diode during the daytime assuming the on-state voltage of the diode is 0.6V?

P5.6   The I-V characteristic for a PV module under STC is given in Figure 5.21. Making any reasonable assumptions

a. Determine the value of the load resistance that will draw maximum power when it is directly coupled to the module.

b. Assume that the average hour-by-hour insolation available for the module is as shown in Figure 5.22. Draw the hour-by-hour I-V curves for the module assuming all other variables remain unaltered.

c. Using the value of the resistive load calculated in part (a), estimate the total energy in watt-hour delivered to the load in one day.

d. Assume that a MPPT is deployed to improve the efficiency of the above module. Estimate the total energy in watt-hours delivered to the above load in one day (between 8 am and 4 pm).

P5.7   A PV module has the hour by hour I-V curves shown in Figure 5.23. Making any reasonable assumptions

a. Determine the value of the load resistance that would draw maximum power from the PV module at 12:00 hours.

b. Determine the daily energy (from 8 am to 4 pm) in watt-hours supplied to the above load resistance.

c. The above load resistance is now supplied by two such modules connected in series. Determine the daily energy supplied to the load.

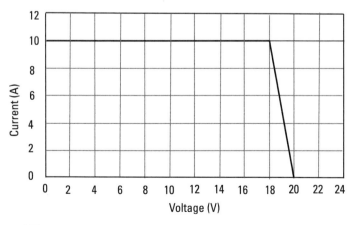

Figure 5.21   I-V curve.

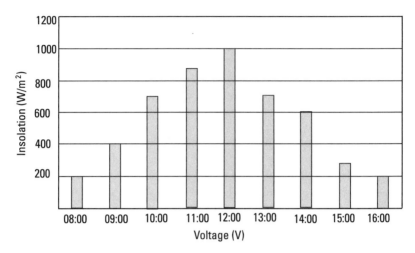

Figure 5.22   Average hourly insolation from 8 am to 4 pm.

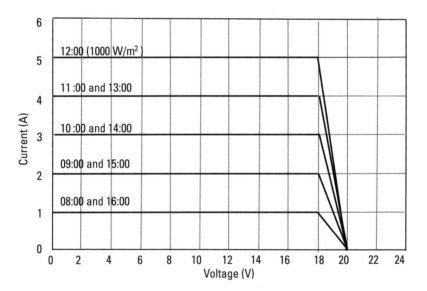

Figure 5.23   Hour by hour insolation curves.

        d. The above load resistance is now supplied by two such modules connected in parallel. Determine the daily energy supplied to the load.

P5.8   A small DC motor is directly coupled to a PV module. The I-V curve of the motor and the hour-by-hour I-V curves of the PV module are shown on the same axes in Figure 5.24. Once the motor is running, it requires a minimum of 12V to keep spinning. What is

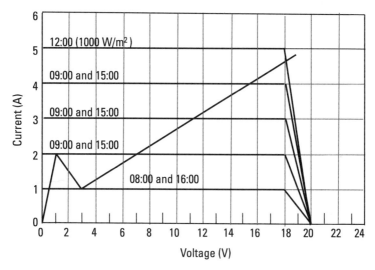

Figure 5.24 Motor I-V and hourly insolation curves.

the earliest time in the morning the motor can be started and what time it will stop running?

P5.9 A permanent magnet DC motor has an armature resistance of 1.2 ohms and spins at its rated speed of 100 rad/s when connected to 18V DC supply drawing a current of 2A.

a. Determine the motor back EMF.

b. Calculate the mechanical power and torque developed by the motor.

c. The motor is to be coupled directly to an 18V PV module whose output current is given as a function of insolation as $G(w/m^2) = 0.008G$. Making any reasonable assumptions, determine the level of insolation needed to keep the motor running at its rated speed.

References

[1] Best, R. E., *Phase-Locked Loops: Design, Simulation, and Applications*, 6 Ed., New York: McGraw-Hill, 2007.

[2] Riley. K. F., and M. P. B. S. J. Hobson, *Mathematical Methods for Physics and Engineering: A Comprehensive Guide*, Cambridge, UK: Cambridge University Press, 2006.

[3] Energy Networks Association, *Engineering Recommendation G83*, London: Operations Directorate of Energy Networks Association, 2012.

[4] IEEE, *IEEE 1547 Standard for Interconnection and Interoperability of Distributed Energy Resources with Associated Electric Power Systems Interfaces*, New York: IEEE, 2018.

[5]   Anani, N., O. Al-kharji, M. AlQutayri, and S. Alaraji, "Synchronization of a Renewable Energy Inverter with the Grid," *Journal of Renewable and Sustainable Energy*, Vol. 4, No. 4, 2012, p. 043103.

[6]   Mohan, N., T. M. Underland, and W. P. Robbins, *Power Electronics: Converters, Applications, and Design*, Washington, DC: John Wiley & Sons, 2002.

[7]   Elobaid, L. M., A. K. Abdelsalam, and E. E. Zakzouk, "Artificial Neural Network-Based Photovoltaic Maximum Power Point Tracking Techniques: A Survey," *IEEE Transactions on Power Electronics*, Vol. 9, No. 8, 2015, pp. 1043–1063.

[8]   Kottas, T. L., Y. S. Boutalis, and A. D. Karlis, "New Maximum Power Point Tracker for PV Arrays Using Fuzzy Controller in Close Cooperation with Fuzzy Cognitive Networks," *IEEE Transactions on Energy Conversion*, Vol. 21, No. 3, 2006, pp. 793–803.

[9]   Chen, L., C. Tsai, Y. Lin, and Y. Lai, "A Biological Swarm Chasing Algorithm for Tracking the PV Maximum Power Point," *IEEE Transactions on Energy Conversion*, Vol. 25, No. 2, 2010, pp. 484–493.

[10] Sera, D., L. Mathe, T. Kerekes, S. V. Spataru, and R. Teodorescu, "On the Perturb-and-Observe and Incremental Conductance MPPT Methods for PV Systems," *IEEE Journal of Photovoltaics*, Vol. 3, No. 3, 2013, pp. 1070–1078.

[11] Hadjipaschalis, I., A. Poullikkas, and V. Efthimiou, "Overview of Current and Future Energy Storage Technologies for Electric Power Applications," *Renewable and Sustainable Energy Reviews*, Vol. 13, No. 6-7, 2009, pp. 1513–1522.

[12] Aurbach, D., Y. Gofer, Z. Lu, et al., "A Short Review on the Comparison Between Li Battery Systems and Rechargeable Magnesium Battery Technology," *Journal of Power Sources*, Vol. 97, 2001, pp. 28–32.

[13] Masters, G. M., *Renewable and Efficient Electric Power Systems*, New York: John Wiley & Sons, 2004.

[14] Guena, T., and P. Leblanc, "How Depth of Discharge Affects the Cycle Life of Lithium-Metal-Polymer Batteries," *INTELEC 06 - Twenty-Eighth International Telecommunications Energy Conference*, Providence, RI, 2006.

6

Hydroelectricity

6.1 Learning Outcomes

After actively engaging with the material in this chapter, you should be able to

1. Describe the magnitude of the global hydro resource;
2. Explain the physical principle of operation of a hydroelectric power plant;
3. State the main elements of a hydroelectric plant;
4. Discuss the main types of hydroelectric plants;
5. Describe the characteristics of the main types of water turbines;
6. Calculate the energy available from a hydro resource;
7. Describe the principle of pumped storage plants;
8. Discuss the advantages and disadvantages of hydroelectricity.

6.2 Overview

This chapter is an introduction to hydroelectricity (i.e., the conversion of energy in moving water into electrical energy). It starts by introducing waterpower and the physical principle of extracting energy from moving water in a hydroelectric power plant. It explains the classification of hydropower plants based on their rated capacity and water head, along with the characteristics of three different types of water turbines. The chapter also presents a simple method for calculating energy available for extraction from a hydro resource. Global hydro resources and installed capacities are also briefly presented. Finally, a set of problems are provided at the end of the chapter to consolidate understanding of the material presented in the chapter.

6.3   Introduction

The term hydroelectricity refers to electrical energy that is generated by extracting energy from moving water. Water energy is another form of renewable energy that is solar in principle. It is the sun that heats water in oceans, rivers, and lakes, turning it into water vapor that rises into the atmosphere. When the water vapor cools in the atmosphere, it turns back into liquid and falls on ground, rivers, and oceans. Water energy is therefore part of the natural water cycle, and consequently, it is a renewable source of energy. Moving water in rivers and streams has kinetic energy and water held at a height has stored potential energy. Direct use of water energy has been used for many centuries; for example, watermills have been used for grinding corn and flowing streams have been used to raise water for irrigation and drinking. Indirect use of waterpower (i.e., for generating electricity) has been used for over a century and currently it is a major source of renewable energy, contributing to about 20% of the world's primary energy consumption. The kinetic energy in moving water, or the potential energy stored in water at a height, can be converted into electricity using a turbogenerator (i.e., a water turbine driving an electric generator). A water turbine is a mechanical machine in which the kinetic energy in a moving water is converted to rotational mechanical energy by the impulse or reaction of the water with a series of buckets, paddles, or blades, which are mechanically fitted around the circumference of a wheel or a shaft. The complete system that extracts energy from moving water and converts it into electricity is called a hydroelectric power plant.

6.4   Essential Elements of a Hydroelectric Power Plant

Dams can be constructed to accumulate and store water in lakes, reservoirs, and rivers at some height for later release through a penstock, which is basically a pipe for conducting water to the turbine, as depicted in Figure 6.1. The dam has sluice gates that allow diversion of water as required. This is because in addition to using water for generating electricity, it can also be diverted for other uses, such as for drinking and irrigation. In the reservoir the dam allows water to be stored at some height, H (m), which gives the stored water potential energy.

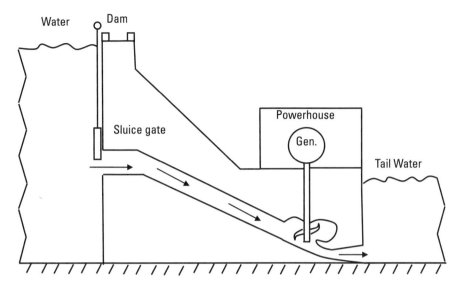

Figure 6.1   Basic elements of a hydro power plant.

Water can be allowed to flow through the sluice gates and the penstock to turn a turbogenerator to produce electricity. The powerhouse houses the electrical generator, which is coupled to the turbine using a driveshaft. Further, in some hydropower plants, at times when hydropower generation is not needed, power from the grid can be used to drive the generator as a motor to pump the tail water up into the reservoir. This water can later be used to generate hydroelectricity during high load demand.

6.5   Classifications of Hydroelectric Plants by Rated Capacity

Hydroelectric power plants or stations are classified according to their capacity (i.e., the rated power output or their effective head, H (m)). The effective head is the vertical difference between the head water level (reservoir) and the tail water (downstream) level. In general, hydropower plants with less than 10-MW capacities are considered small plants. However, there is no universal agreement on this classification; for example, in the United States the above figure is 30 MW and in China it is 50 MW. Microhydroelectric plants have capacities from a few hundred watts to a few hundred kilowatts. Regardless of the rated capacity, all hydropower stations work on the same physical principle, which uses the energy in moving water to rotate a turbine that turns a generator. They only differ in their size and power output. However, there are reasons for providing statistics for hydropower plants according to their capacities. This is because in some countries, there are incentives for renewable energy plants, and only small hydro plants count as renewables. Understandably, if we count large hydropower plants as renewables, they will mask out the growth of other renewable sources, such as wind and solar. Therefore, small hydroelectric plants are counted and reported as renewables, while large plants greater than 50 MW are not counted as an addition to renewable energy generation [1]. There is also the debate that large hydro plants require large reservoirs and large civil engineering constructions, which can have negative social and environmental impacts, and thus fail to comply with the requirement of a sustainable source of energy stated in Chapter 1 (a sustainable source of energy should not have any adverse effects on the environment or wildlife, and should not lead to any social injustice).

6.6   Classification of Hydropower Plants by Effective Head of Water

In a hydroelectric power plant, the output power depends on the head, $H(m)$ and the flow rate $Q(m^3/s)$. Therefore, the amount of energy that can be generated is proportional to the head and the flow rate. A low head hydroelectric plant uses a relatively large volume of a slowly moving river, Figure 6.2. The barrage is constructed to intercept the watercourse to increase the volume of water and improve the flow rate.

A medium head plant, shown in Figure 6.3, is typical of the large hydroelectric plants with a dam at a narrow point in a river valley. The large reservoir behind the dam provides enough storage to meet demand.

On the other hand, a high head plant uses a lower volume of water and high-speed water flowing from a dam, as seen in Figure 6.4. Generally, a high head implies an effective head of more than 100m and low water head less than 10m.

A low head plant uses relatively large amounts of water and usually has low capacity and requires large civil engineering construction to accommodate the large amounts of water. The higher the head, the higher is the water pressure on the hydro turbine and the more power it will generate. In a high head plant, the entire reservoir is well above the outflow, and the water flows through a long

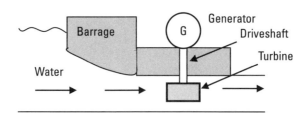

Figure 6.2   Low head hydroelectric power plant.

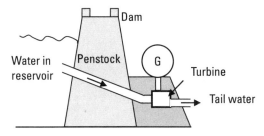

Figure 6.3 Medium head hydroelectric plant.

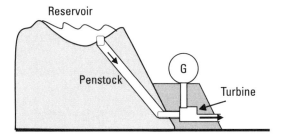

Figure 6.4 High head hydroelectric plant.

penstock carrying the flow to the turbine. Higher heads are preferable to lower heads, not only because they can generate more power, but also because higher heads mean higher flow rates, and consequently, smaller and hence cheaper turbines and generators, and a more compact powerhouse.

There is another type of hydroelectric plant called run-of-the-river [2]. These plants harvest the energy from naturally moving water (e.g., in rivers and streams) without requiring large dams and reservoirs, although small dams are sometimes constructed to improve efficiency. The main differences between run-of-the-river plants and other types is that they use the natural flow rate of water, caused by rainfall or melting snow, to generate electricity instead of the energy in water falling from high reservoirs. They are primarily used in locations where there is little or no space for water storage. Run-of-the-river plants require a site where there is a substantial flow rate, usually from rainfall, and enough water speed. The main drawback of this type of plant is reliability because its output depends on weather conditions. The classification of hydroelectric power plants based on the size of the head is important because the head size determines the type of the turbine that can be used for a given head size.

6.7   Calculation of Hydropower

The important parameters for energy calculations in a hydro system are the effective head H (m) and the flow rate Q (m³/s) (i.e., the number of cubic meters of water passing through the plant per second). These are related to the power delivered by the water P as explained below.

When a mass m (kg) is raised through a height H (m) above the ground, it has a stored potential stored energy E_P (J) given by

$$E_\mathrm{P} = mgH \tag{6.1}$$

where g (m/s²) is the acceleration due to gravity. In the context of hydropower, we are not normally interested in the mass of water; instead we are interested in the volume V(m³) of water because this is used to specify the capacity of a reservoir. If ρ(kg/m³) (the Greek letter rho) is the density of pure water, the mass is

$$m = \rho V \tag{6.2}$$

Hence, we can write for the stored potential energy at height H (m)

$$E_\mathrm{P} = \rho gVH \tag{6.3}$$

We are interested in the rate at which this energy is delivered (i.e., the power P (W) delivered to the plant0. To do this, we divide energy by time t (s)

$$P = \rho g \frac{V}{t} H \tag{6.4}$$

That is, the input power is given by

$$P = \rho gQH \tag{6.5}$$

where Q is the water flow rate in cubic meters per second (m³/s). In a real system, as water falls down a pipe, it will lose some energy due to friction with the pipe walls and due to some turbulent flow. Therefore, the effective head is always less than the actual gross head. In addition, there are other losses in the plant; for example, in the turbine and generator. As a result, the output power from a plant will always be less than the input power. If the overall efficiency of a plant is η (the Greek letter eta), then the output power in watts becomes

$$P_o = \eta \rho gQH \tag{6.6}$$

We can use the approximate value of density for pure water (i.e., $\rho = 1000$ (kg/m^3) and the approximate value of the gravitational acceleration g = 10(m/s^2) and write for the output power in watts :

$$P_o = 10000\eta QH \tag{6.7}$$

Or, in kilowatts, the output power is

$$P_o \text{ (kW)} = 10\eta QH \tag{6.8}$$

Example 6.1

Consider two hydroelectric plants, each with an efficiency of 80%. The first plant has an effective head of 30m and a flow rate of 3000 liters per minute. The second plant has an effective head of 120m and a flow rate of 4000 cubic meters per second. Determine the output power for each plant.

Solution:

For the first plant the flow rate is

$$Q = \frac{3000 \div 10^3}{60 \ s} = 0.05 \ \text{(m}^3\text{/s)}$$
$$P_{O1} \text{ (kW)} = 10\eta QH = 10(0.80)(0.05)30 = 12 \ \text{kW}$$

For the second plant

$$P_{O2} \text{ (kW)} = 10\eta Q_2 H_2$$
$$= 10(0.80)(4000)(120)$$
$$= 3.84 \times 10^6 \ \text{kW}$$

The water head and the power rating of a hydropower plant are major factors in determining the most appropriate type of turbine for the plant. A turbine includes of a set of blades fitted around the circumference of a shaft, which are specially curved to deflect water in such a way that most of the water energy is converted to mechanical rotation energy that rotates the shaft. The blades and the shaft along with their support structure constitute the turbine runner. Depending on the type of turbine, water is directed on to the runner by channels and guide vanes or through a jet. There are many types of turbines with runner diameter ranging from 20 cm to 10m. However, based on their principles of operation, there are three main types that are commonly used in hydroelectric plants: Francis, Kaplan, and Pelton turbines.

6.8 Pelton Turbine

The Pelton turbine, or Pelton wheel, shown in Figure 6.5, is an impulse type of turbine. An impulse turbine is driven by a high-velocity jet of water coming out of a nozzle and striking buckets fitted around the circumference of a wheel. The resulting impulse force spins the wheel and removes most of the kinetic energy from the water. Before reaching the turbine wheel, the nozzle converts the pressure in the water head into kinetic energy (i.e., fast-moving water). The Pelton turbine was invented in the United States by Lester Pelton in the 1870s, and it is most suitable for hydropower generation when the available water source has a high head, typically greater than 300m, and low flow rates [3]. The wheel of a Pelton turbine consists of several spoon-shaped double cups or buckets mounted on the rim, as shown in Figure 6.5. The high pressure caused by the

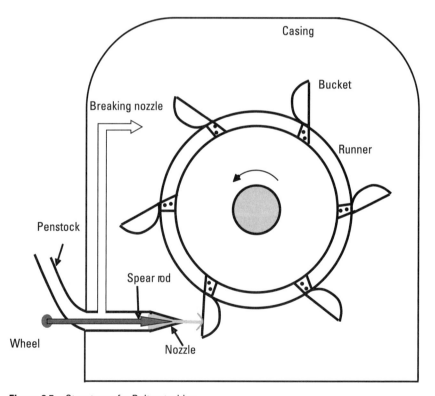

Figure 6.5   Structure of a Pelton turbine.

high head enables the nozzle to convert the stored potential energy into kinetic energy. The Pelton turbine converts this kinetic energy into rotational mechanical energy that turns a generator.

The flow rate can be controlled from zero to maximum by moving the spear rod inside the nozzle, also known as the needle. For zero flow, the nozzle can be used to direct the flow away from the wheel. Under optimum conditions, almost all the kinetic energy in the water is delivered to the turbine. Maximum power extraction in a Pelton turbine is obtained when the product, P, of the of the impulse force and cup velocity is maximum

$$P = F_{\text{impulse}} \times V_{\text{cup}} \qquad (6.9)$$

If the cup velocity is zero, the efficiency is zero even if the impulse force is very large. Similarly, if the cup velocity is the same as the water jet velocity, the efficiency is also zero since it will be very difficult for the jet to hit a cup (i.e., the impulse force is zero). The optimum speed for a Pelton turbine is found to occur when the cup speed is half the speed of the jet.

6.8.1 Governing the Pelton Turbine

To stop the generator, the nozzle can be diverted away from the wheel; however, since the wheel is very heavy, it will continue running for a long time due to its inertia. To stop the runner in a short time, the upper nozzle (the breaking nozzle), shown in Figure 6.5, can be used to direct a water jet on the back of the buckets to stop it more quickly. Under normal operating conditions, since the water head is fixed, control of the output power from the generator driven by the turbine is accomplished by moving the spearhead (inside the outlet nozzle) back and forth by a servomechanism in order to control the diameter of the water jet. The reader is referred to [3] for more details on the Pelton turbine.

6.8.2 Input Power

In practice, water leaves the jet at a lower speed than that it would have gained through a free fall from the height of the effective head, H (m), due to losses in the jet (e.g., due to friction). In the following, we will ignore these losses and assume that water leaves the jet at the same speed it would have gained in a free fall through the effective head under the force of gravity. The potential energy given up by mass m (kg) of water falling from height H (m) is $E_P = mgH$, and the kinetic energy of mass m (kg) of water moving with velocity v (m/s) is $mv^2/2$. If the conversion from potential to kinetic energy is lossless, then we can write

$$\frac{1}{2}mv^2 = mgH \tag{6.10}$$

The speed of the water coming out of the nozzle is therefore given by

$$v = \sqrt{2gH} \tag{6.11}$$

If this water flows as a jet of circular cross-sectional area A (m²), the volume of water flowing out in each second is equal to the area A (m²) times the speed v (m/s). Hence, the volume flow rate for an effective head H (m) is given by

$$Q = A \times \sqrt{2gH} \tag{6.12}$$

The input power, in watts, to a Pelton turbine is

$$P = \rho Q g H \tag{6.13}$$

Substituting for Q from (6.12) the input power becomes

$$P = A\sqrt{2gH}\ \rho g H \tag{6.14}$$

Using $\rho = 1000$ kg/m³ and g=10 m/s²

$$P = 10000 A\sqrt{20}\ \sqrt{H^3} \tag{6.15}$$

Or power in kilowatts is

$$P(\text{kW}) = 10A(4.4721)\sqrt{H^3} \tag{6.16}$$

Hence, the power in kilowatts is approximately

$$P(\text{kW}) = 45A\sqrt{H^3} \tag{6.17}$$

Some Pelton turbines can be fitted with more than one jet spaced around the runner to increase the output without increasing the size of the turbine. If the number of jets is N, then the power input becomes

$$P(\text{kW}) = 45NA\sqrt{H^3}\ \ (kW) \tag{6.18}$$

This equation tells us that the input power to a Pelton turbine varies with the square root of the head cubed. That is, if the head is doubled, the input power will increase by a factor of 2.82 (=$2^{3/2}$).

6.9 Kaplan Turbine

A propeller or axial turbine is one in which the flow of water is parallel to the shaft, while in a radial turbine the flow of water is radial to the shaft. An axial turbine is a reaction type turbine that develops torque by reacting to the pressure or weight of water. The operation of a reaction turbine can be explained using Newton's third law of motion; for every action there is an equal and opposite reaction. A basic propeller turbine is shown in Figure 6.6. The area of water swept by the turbine is determined by the area of the blades. Hence, designers can design the blades to sweep any required size of area. This makes propeller turbines suitable for large volume flows and low heads that are typically 2 to 25m and 50 to 800 m³/s, respectively. One of their important advantages over radial turbines is the fact that it is easy to improve their efficiency by simply varying the angle of the blades to suit variations in the power demand [4]. Axial flow turbines with this facility are known as Kaplan turbines. One important feature of the Kaplan turbine is that for maximum efficiency, the speed of the blades is considerably greater than the water speed and could be as much as twice as fast. This is useful for getting higher speeds for generators when the water speed is relatively low. Hence, Kaplan turbines are most popular for small

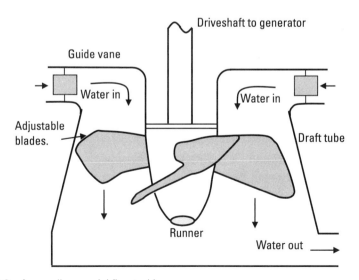

Figure 6.6 A propeller or axial flow turbine.

water heads and high flow rates. The reader is referred to [4] for more details on the Kaplan turbine.

In a Kaplan turbine (see Figure 6.7) flow is entered through a spiral casing whose decreasing area ensures that water enters the central portion almost at uniform velocity throughout the perimeter. Water passes the guide vanes then through to the runner and finally leaves through a draft tube [4].

Since power demand may fluctuate over time, a governing control system is used to control the position of the guide vanes in order to control the water flow to keep it synchronized with demand. In addition, the blades can be twisted to optimize the speed of the runner under varying flow speeds.

One important difference between the Pelton turbine and reaction turbine is that while reaction turbines are fully submerged in water and require a pressure difference across the runner, impulse turbines operate in air under normal atmospheric pressure.

6.10   Francis Turbine

Francis turbines are the most prevalent type in medium- and large-scale hydroelectric plants [5]. They can work efficiently under a wide range of operating conditions. The most significant part of a Francis turbine is its runner, which is fitted with some complex blades designed to maximize its efficiency, as shown in Figure 6.8. Water enters the runner radially and leaves it axially. For this

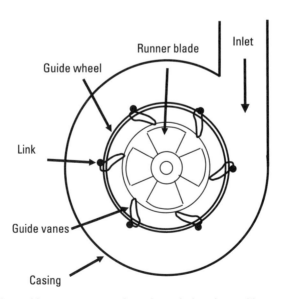

Figure 6.7 Kaplan turbine: water enters through a spiral casing and leaves through a draft tube.

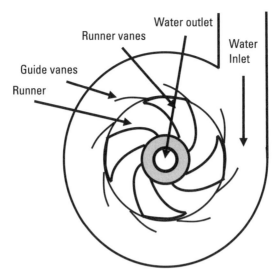

Figure 6.8 A Francis turbine.

reason, Francis turbines are also known as mixed-flow turbines. The blades are specially shaped so that they have an airfoil cross section, so consequently when water flows over the airfoil section, a low pressure is induced to one side and a high pressure is induced on the other side producing a lift force. In addition, the blades have a bucket shape at the outlet side. Hence, when water hits this area it produces an impulse force before leaving the runner. Both the impulse and the lift force will force the runner to rotate. Therefore, although a Francis turbine is classified as a reaction turbine, a fraction of the force comes from an impulse force [5]. The reader is referred to [5] for more details on the Francis turbine.

6.11 Tubular Turbine

A tubular turbine is suitable for run-of-the-river hydro plants. It is an axial pro-peller type that is suitable for low pressure and very low water heads, and doesn't require a penstock. The generator may be enclosed in a watertight bulb-like structure, as shown in Figure 6.9; hence, in this case it is called a bulb turbine. Alternatively, the generator may be situated on the riverbed and coupled to the turbine using a long shaft. In this case it is referred to as a tube turbine.

6.12 Load Curve

The load curve is a graphical representation of the variation of electrical load, usually in MW or GW in a proper time sequence over a period. A daily load

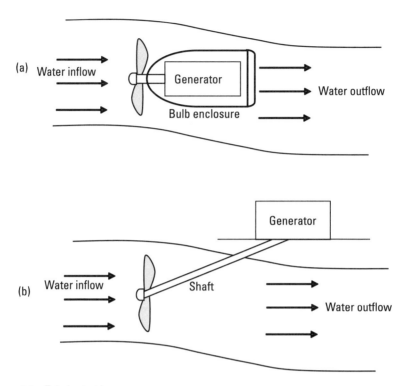

Figure 6.9 Tubular turbines.

curve shows variation of the load on a power station over a 24-hour period. The load curve, also known as the load profile, varies from day to day and from season to season. A typical daily load curve may look something like that shown in Figure 6.10. The load curve provides a variety of information, such as the variation of the load during different hours of the day and the base load that a power station must maintain during the whole day. In addition, the load curve shows the peak load, which determines the maximum load demand on the power station. The area under the curve represents the total energy generated under any required period.

6.13 Pumped Storage

The load curve can be unpredictable and sudden increases and dips in demand are not uncommon. This means that a power station must respond to sudden changes in load demand. This is not possible for certain power plants; for example, the output of a nuclear reactor cannot be adjusted on an hourly basis

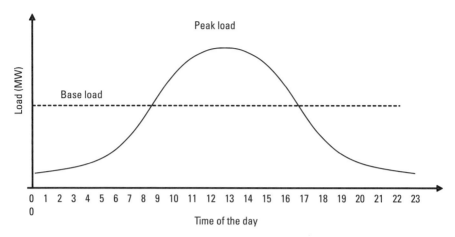

Figure 6.10 A typical daily load curve.

to follow variations in load demand. That is, if the load increases suddenly, the output of a nuclear power plant cannot be increased to meet the new load. Similarly, if the load suddenly drops, the output of the station cannot be reduced to match the new load demand. Further, some plants are most efficient when running at full capacity. Therefore, the solution is to find some means of storing excess electrical energy when load is low and be able to retrieve the stored energy fast enough to meet a sudden increase in load.

In addition to nuclear power plants, increased use of renewables, such as wind and photovoltaics, have been the driving force for recent increased interest in storage of electrical energy. The output of renewable energy plants such as PV and wind depend on weather conditions that can be constantly varying; therefore some means of backup is required. Small amounts of electrical energy may be stored in various systems such as batteries or compressed air, but storing large amounts is, at least at present, only viable using water. The principle is simple. Electrical energy is converted into gravitational potential energy when the water is pumped up from a lower reservoir to an upper one, and the process is reversed when it is released to run back down, driving a turbogenerator on the way (Figure 6.11). This method of storage requires a generator that can be run backwards as an electric motor and a turbine that can run in both directions (i.e., either extracting energy from the water or delivering energy to the water). The advantage of using hydropower stations for storing electrical energy is that their response time is very short, typically around 10 minutes. Pumped storage has become increasingly important, with installed capacity worldwide having grown from 78 GW in 2005 to reach 153 GW in 2017, and during the year 2017, 3.2 GW of pumped storage capacity was installed [6].

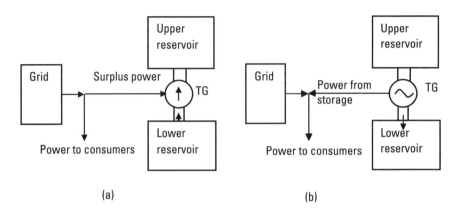

Figure 6.11 Pumped storage system at time of low demand (a) and high demand (b).

6.14 Hydro as an Element in a Power System

In addition to environmental and economic factors, a major factor to consider when designing a new power plant, is how easy it is to integrate the new plant with an existing power system. From an operational point of view, this integration depends on several factors, such ss availability of the plant, its input energy, and its response time. Hydropower plants, particularly large ones with pumped storage, can easily cope with variations in rainfall and their availability is not normally hindered by occasional dry seasons. In addition, hydropower plants have a very fast response time and their output power can be easily and rapidly adjusted to match variation in demand. We have seen that the output of a Pelton turbine, for example, can be adjusted by simply moving a spearhead in its nozzle.

As we have seen earlier, the daily load curve shows that the demand consists of a base load that a power station has to constantly maintain, and a peak load that is typically required for about 20% of the time. When planning a new hydropower plant, we have the choice of designing it for supplying either the base, the peak demand, or both.

Example 6.2

A proposed hydropower plant is to provide an annual energy output of 400 GWh. Determine the required rated capacity if the plant is to supply (a) base load only and (b) peak load with 20% capacity factor (CF).
    Solution:

$$\text{Rated capacity} = \frac{\text{actual annual power generated}}{8760 \ (\text{hours}) \times \text{CF}}$$

for a base load the capacity factor is 100%. Hence, the required rated capacity is

$$\text{Capacity} = \frac{400 \times 10^9}{8760 \text{ (hours)} \times 1} = 46 \text{ MW}$$

For the peak load with a capacity factor of 20%, the required rated capacity is

$$\text{Capacity} = \frac{400 \times 10^9}{8760 \text{ (hours)} \times 0.2} = 230 \text{ MW}$$

Clearly, designing a plant for a peak load requires a much higher capacity.

6.15   Resource

The hydropower resource is clearly the power that can be extracted from water in a hydro plant due to the presence of water head or naturally moving water. It is estimated that about a quarter of the 5.4 million EJ of solar energy reaching the earth each year is used in the evaporation of water. That is, the water vapor in the atmosphere represents a huge amount of energy that is stored and constantly replaced. However, a substantial amount of this stored energy is wasted as heat in the atmosphere as vapor condenses into water. The fraction of energy that reaches the earth as rain falls is estimated at around 200,000 TWh per year and only about 40000 TWh/y of this energy in the form of flowing water is regarded as the world's resource of hydro energy [1].

However, it is not always technically possible to build a hydro plant on every possible site, therefore the actual usable hydro resource is reduced to about 16,000 TWh/y. This is known as the technical potential [1]. Further, even this technical potential cannot be completely used for economic reasons. For example, it is not economically viable to construct a hydro plant at a given suitable site if the price of energy from other fuels, such as gas or oil, are very competitive. This again reduces the technical potential to what we call the economic potential for hydroelectricity. Regional resources of hydro energy vary considerably; for example, the technical potential in Asia is about 5,590 TWh/y while in the Middle East is only about 280 TWh/y. However, the extent to which this hydro is exploited is another matter. In 2016, Asia had a share of 42.4% compared to the total world's total hydro generation, while Europe 24.4% and North America had 16.1%[7]. According to the 2018 status report of the International Hydropower Association, in 2017 the global installed capacity of hydropower plants climbed to 1267 GW and a record 4,185 terawatt hour, of electricity was generated, saving the earth from 4 billion tonnes of greenhouse gases [6]. In terms of the installed capacity, in 2017 China was leading the world with 9.12 GW, followed by Brazil with 3.38 GW, and India with 1.91 GW [6].

6.16   Advantages and Disadvantages of Hydropower

Hydropower uses water as fuel, so it is a clean source of energy and does not generate greenhouse gases. It is domestic so it can be localized and it supports the current trend of distributed generation. It can also be readily integrated with an existing power system and has a fast response time that allows it to respond to hourly variations in load demand. Hydropower remains the only economically viable method of storing electrical energy using pumped storage. In addition, hydropower plants with large reservoirs can prevent floods and help mitigate draught by providing water for irrigation during dry seasons. However, there are some disadvantages of hydroelectricity: it can lead to causing damage to wildlife and lead to social injustice if people have to give up their land and or homes for building large dams.

6.17   Summary

Hydroelectricity is a renewable energy source that extracts the energy in moving water and converts it into electrical energy. The output of a hydropower plant is proportional to the head and water flow rate. Classifying hydro plants according to the size of their head is important as this decides the type of turbine to use. We have looked at the three main types of turbines. The Pelton turbine is an impulse type of turbine that is suitable for high heads and low flow rates, while the Kaplan turbine is a reaction type turbine that uses the pressure of water, and hence is suitable for low heads and large flow rates. The Francis turbine is a reaction turbine used for medium heads. The two major advantages of hydroelectricity are: (1) it supports storage of large amounts of electrical energy in the form of potential energy using pumped storage, and (2) it can be easily integrated with an existing power system to supply peak loads by virtue of its fast response.

6.18   Problems

P6.1   a. Using your own words and with the aid of a diagram explain the principle of electrical power generation in a hydroelectric plant.

b. Discuss the classification of hydroelectric power plants according to their rated capacity.

c. Large hydroelectric plants are not classified as renewables. Discuss this statement.

d. Explain the classification of hydroelectric plants according to their head and the significance of this classification.

e. Explain the difference between a reaction and an impulse turbine.

f. Explain the reason for using a Pelton turbine for high head hydro-power plants.

g. Explain what is meant by the load curve and its significance.

h. Discuss two reasons that make storage of electrical energy in large quantities important and discuss how pumped storage can be utilized for this purpose.

P6.2 A hydroelectric power plant has an average head of 50m and a flow rate of 60 cubic meters per second. It is output capacity is 25 MW. Determine the efficiency of the plant.

P6.3 A 36-MW hydroelectric power plant has an efficiency of 75% and a flow rate of 30 cubic meters per second. Determine its head.

P6.4 A pumped storage hydroelectric plant has a high-level reservoir with storage capacity of 5 million cubic meters of water at an effective head of 300 meters. Determine

a. The maximum energy stored in kWh;

b. The volume of water raised in one hour by a 100-MW machine running at full capacity as a pump at 90% efficiency;

c. The average area of the reservoir.

P6.5 A hydropower plant has an effective head of 300 meters and a water flow of 20 cubic meters per second.

a. Calculate the total power input to the turbine.

b. The water reaches the turbine through a penstock with a cross-sectional area of 2.5 square meters. Making any reasonable assumptions, determine the speed of the water.

c. If the efficiency of the plant is 70%, determine the power output.

P6.6 A hydropower plant has an effective head of 300 meters and uses a 20-MW generator driven by a Pelton turbine whose efficiency is 90%.

a. Calculate the required flow rate of water to run the generator at its full capacity;

   b. Calculate the speed of water through the jet;

   c. Calculate the cross-sectional area of the jet and its diameter.

P6.7   A 150-MW hydroelectric plant has an annual average production of 800 GWh. Calculate its capacity factor of the plant.

References

[1]   Boyle, G. (ed.), *Renewable Energy Power for a Sustainable Future*, Oxford, UK: Oxford University Press with The Open University, 2012.

[2]   Green Energy Futures, "Run-of-River Hydroelectric Power," August 28, 2015, http://www.greenenergyfutures.ca/episode/30-how-it-works-run-river-hydro-electric-power.

[3]   Mekanizmalar, "Pelton Turbine," Mekanizmalar, October 11, 2013, https://www.youtube.com/watch?v=qbyL--6q7_4.

[4]   Learn Engineering, "Kaplan Turbine," Learn Engineering, May 2, 2013, https://www.youtube.com/watch?v=0p03UTgpnDU.

[5]   Learn Engineering, "Francis Turbine," Learn Engineering, July 24, 2103, https://www.youtube.com/watch?v=3BCiFeykRzo.

[6]   International Hydropower Association, "2018 Hydropower Status Report," IHA, London, 2018.

[7]   World Energy Council, "Energy Resources- Hydropower," 2016, https://www.worldenergy.org/data/resources/resource/hydropower/.

7

Tidal Energy

7.1 Learning Outcomes

After actively engaging with the material in this chapter, you should be able to

1. Describe the basic structure of tidal barrage systems;
2. Describe the basic structure of tidal stream systems;
3. Explain the patterns of electricity generation over the tidal cycle;
4. Calculate power and energy available in a tidal stream;
5. Calculate power and energy available in a barrage system;
6. Discuss the main advantages and disadvantages of tidal energy.

7.2 Overview

In this chapter we will consider the ways in which tidal energy can be harvested and converted to electrical energy. We will study the single- and double-basin schemes for generating tidal electricity and consider the variation of the generated electricity during the course of a tidal cycle. In addition, we will use a simplified mathematical approach to derive expressions for energy and power in tidal plants. Several problems are presented at the end of the chapter to consolidate understanding of the material.

7.3   Introduction

The tides of seawater (i.e., the rise and fall of the seawater) represents a vast amount of energy that can be used to generate electricity. Unlike hydropower, which is caused by the hydrological cycle that is driven by the sun's energy and is usually harvested using large dams, tidal energy is the result of the gravitational attraction of the sun and moon on the oceans of the earth as was explained by Newton in 1687 [1]. However, the effect of the moon is much larger than the effect of the sun. Tidal power relies on the upstream flows (called the flood) and downstream flows (called the ebb) in river estuaries and lower reaches of some rivers and seas. Tidal energy can be harvested using tidal barrages or lagoon plants and tidal stream plants [2]. Tidal barrages and lagoons capture seawater behind a barrage or in a specially constructed lagoon at high tide, and then as the seawater falls, the trapped water is released through a turbine to generate electricity. Tidal stream systems use the horizontal movements of water during a tidal cycle to turn a turbine.

The gravitational interaction between the earth and the moon and their mutual rotation is the primary cause of tides that takes place twice during a lunar day. That is the cycle of rise and fall repeats twice every lunar day, which is 24.8 hours (i.e., the cycle of rise and fall has a period of 12.4 hours or 12 hours and 24 minutes). At any given point on Earth there will be two high tides and two low tides; that is, two tidal cycles per lunar day (24.8 hours) occurring 12.4 hours apart. The effect of the sun on the tidal phenomenon is very small compared to that of the moon. The solar tide either enhances the lunar tide or cancels a portion of it, depending on the relative positions of the sun, the moon, and the earth. Hence, when we talk about the tidal cycle we will always imply the lunar tidal cycle.

7.4   The Nature of Tidal Energy

The tidal rise and fall cycle is predictable and is twice during a lunar day (i.e., with 12.4 hours between each cycle). A consequence of the tidal rise and fall of seawater is the formation of upstream floods and downstream ebbs in river estuaries and the lower reaches of rivers and seas [3]. When seawater level rises upward, its vertical motion causes tidal currents (upstream or flood stream) to move horizontally toward the land away from the sea, as illustrated in Figure 7.1. When seawater falls vertically downward, it causes tidal currents to move horizontally toward the sea and away from the land (downstream or Ebb stream), as shown in Figure 7.2.

The difference between the highest level of water attained during a flood and the lowest level after an ebb defines the tidal range R(m). There are two

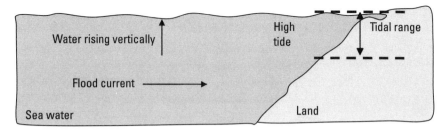

Figure 7.1   Tidal flood.

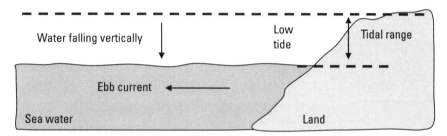

Figure 7.2   Tidal ebb.

different approaches to harvesting tidal energy. The first one is to exploit the periodic rise and fall of the seawater level using barrages, while the second is to harvest the tidal energy in the tidal currents, known as tidal streams, in a manner similar to harvesting wind energy (e.g., using horizontal wind turbines [3]).

7.5   Tidal Barrage Power Generation

A location with suitable tidal range, such as an estuary of a river, can be artificially enclosed with a barrier formally known as a barrage, which has sluice gates and a gated turbine; the whole structure is called a basin. The sluice gates allow water flow across the barrage while the turbine gate allows water to flow through the turbine when it is required to generate electricity. The turbine gate is closed when no power can be or is required to be generated. There are a number of ways of exploiting tidal barrages using either a single or double basin scheme as explained below. Basically, there are three different schemes of generating electricity using a single-basin single-barrage system. They all include the fundamental elements of a barrage, which acts as a barrier for trapping water, a sluice gate, and a gated turbine, as shown in Figure 7.3.

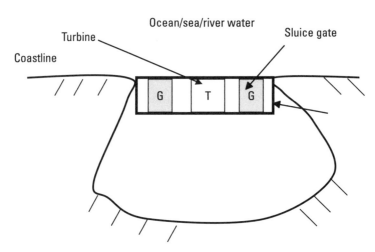

Figure 7.3 Single-basin tidal energy plant.

7.5.1 Single-Basin Ebb Generation

During a tidal rise, the sluice gates are opened and the incoming seawater fills the basin (see Figure 7.3). At the high tide (i.e., once the tide is at its highest level) the sluice gates are closed, the basin is full, and the water is trapped in the basin by the barrage. The water is then held in the basin for approximately 3 hours. During this time there is no flow and no power generation. This is because there is not enough head for generation because the seawater is still high and close to the level of water in the basin.

When the seawater starts to fall and only after it has fallen to approximately half the tidal range, which takes about 3 hours, the turbine gate is opened and the basin water is allowed to flow through the turbine generating power. This will continue until the head becomes too small to generate any power, which usually it takes about 5 hours. The turbine gate is closed and the sluice gates are opened for the next tidal rise to start filling the basin. Clearly, since energy generation is during the ebb, this system is called ebb generation.

7.5.2 Single-Basin Flood Generation

In the case single-basin flood generation, the sluice gates are kept closed during a flood (i.e., tidal rise) until enough head is established outside the barrage, about half the tidal range. Once a suitable head is achieved, the turbine gate is opened and water flows through the turbine into the basin generating electricity. The generation usually lasts for about 5 hours. When the basin water reaches the sea level, the sluice gates are opened and generation stops. This stage normally takes about 3 hours. Then the cycle repeats. Flood generation is less favorable than ebb generation as keeping the tidal basin at a low water level for prolonged period of time could have adverse effects on the environment and on shipping. Further, it is less efficient than ebb generation since the surface area of the basin at high tide is larger than at low tide, so more water is available during the ebb generation.

7.5.3 Single-Basin Two-Way Generation

Single-basin two-way, generation can be obtained during both the ebb and the flood tides. This is achieved by using a two-way turbine or two turbines, one for each tide. However, studies indicate that there is no major advantage in doing this and that the power generation will not be doubled relative to the single-turbine ebb scheme. Economically it is not very attractive because of the extra cost of the two-way turbine. However, it can reduce the periods of time during which there is no power generation and also reduces the peak power, which reduces the rating, and hence, the cost of the generator [4]. A common disadvantage to all single-basin systems is that the power delivered is discontinuous because there is generation only during parts of the tidal cycle and there is no way of adjusting the supply to match the power demand. An improvement to the single-basin scheme is to use a double-basin system, which improves the continuity of power supply.

7.5.4 Double-Basin Scheme

The double-basin system uses two basins and two barrages, as shown in Figure 7.4. One barrage has a sluice gate, one for each basin, and the common barrage between the two basins has a one-way turbine. The system uses the head of water between the upper and lower basins. During the tidal rise period, the sluice gate of the lower basin is closed, while the gate of the upper basin is opened to the sea to fill with water. During the low tide period, the gate of the lower basin will be opened, and therefore the upper basin will maintain a relatively higher head than the lower basin.

At the high tide (i.e., at the highest level of the tidal rise) the sluice gate of the upper basin that is now filled with water is closed and the turbine gate is opened so water flows from the upper to the lower basin, thus generating electricity. The water level in the lower basin will start to rise because its gate is closed. Therefore, the differential head between the water levels in the basins starts to decrease. When the head becomes small enough so that it may stop generation of power, the gate of the lower basin is opened so that the head is increased again. The cycle repeats and if the timing of the opening and closure of gates is chosen appropriately, the output power, though of variable level, it is continuous throughout the tidal cycle.

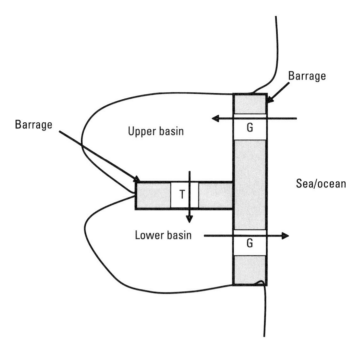

Figure 7.4   Double-basin tidal barrage scheme.

7.6   Power Generation Profile

We will now consider the profile of power generation using tidal barrage energy harnessing. Let us consider a single-basin ebb generation plant. The seawater level during a tidal cycle varies almost sinusoidally with a period of approximately 12 hours and 24 minutes, as shown in Figure 7.5. Starting from $t = 0$ with the tidal rise, the sluice gates are opened and the basin water level very closely follows the seawater level. At the high tide at $t = a$, the basin is full and its level is the same as the sea level; hence, the sluice gates are now closed. Note that during this period there is no head between the water levels in the basin and the sea, $(H = 0)$. For $t>a$, the seawater level starts to fall (start of low tide) but the basin level remains fixed until point $t = b$, which is normally about three hours, when there is enough head to generate electricity. This is typically half the tidal range, so the turbine gate is opened and electricity generation commences. The basin water level starts to fall but at a lower rate than the seawater, which maintains a head that results in electricity generation.

   This generation usually continues for approximately 5 hours. At point $t = c$, the head is maximum and the generated electricity reaches its peak. The tidal rise starts again, but the basin water is still falling, thus the head starts to decrease. At point $t = d$, the head is insufficient for power generation; hence, the turbine gates are closed and the electricity generation stops. Basin water level

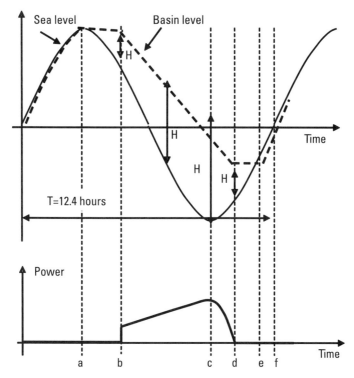

Figure 7.5 Sea and water levels during a tidal cycle (top) and power generation profile (bottom).

remain fixed until point $t = e$ when the sluice gates are opened and the basin starts to fill again following very closely the level of the seawater. From point $t = f$ onward the cycle repeats periodically every 12.4 hours. Therefore, power generation in a tidal cycle is one burst with varying levels due to variations in the head between the sea and the basin water levels. Clearly this is undesirable as it implies that single-basin power plants will have difficulties meeting typical variations in demand for electrical power during the course of a day. This difficulty can be reduced to some extent (i.e., the periods of no-power generation can be reduced using a two-way turbine in the single-basin system), as shown in Figure 7.6. Starting at $t = 0$, during the tidal rise, water is filling the basin through the turbine, thus generating electricity until the head becomes too small at point $t = a$.

The turbine is closed and the water level in the basin remain fixed until $t = b$, when there is enough head again and the turbine is reversed so that it starts generating power during the tidal fall. The power is maximum at the highest head at point $t = c$. At the point $t = e$ the turbine is closed and there is no power generation until the point $t = f$ when the turbine is opened again and the cycle repeats.

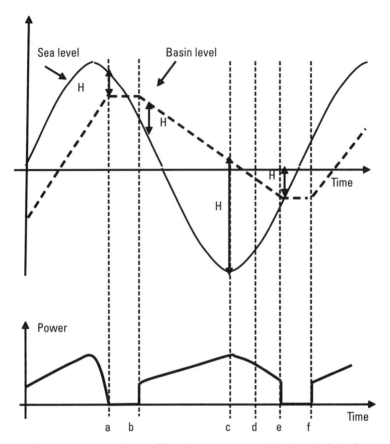

Figure 7.6 The power generation profile using a two-way turbine in a single basin.

However, continuity of the tidal power generation may be achieved using a double-basin system, such as that shown in Figure 7.4, which maintains a differential head between the two basins during the whole tidal cycle, and it is this head used to generate electricity. With this topology it is possible to obtain power without interruptions.

7.7   Tidal Lagoon Energy Harnessing

Tidal lagoons are basically containers of water of circular cross section constructed offshore in open water. They collect and trap water during high tides and release it during low tides through turbines to generate electricity. Lagoons share barrages in the fact that they both use water heads that are caused by a tidal range, and hence, sometimes both are referred to as tidal range systems. However, while a barrage spans an entire river estuary, a lagoon uses a much smaller area that must be chosen with due account taken of the local environment. Lagoons are very expensive to construct, and hence, are less popular than tidal barrage systems.

7.8 Tidal Stream Schemes

Instead of using costly and environmentally invasive barrages located in estuaries of rivers to exploit the vertical rise and fall of tides and the potential energy of heads of the trapped water behind dams and lagoons, it is possible to harness the energy in the horizontal movement of the tides, as shown in Figure 7.7 (i.e., the kinetic energy of the flood and ebb or tidal upstream and downstream). In open seas, however, the speed of the tidal stream (ebb and flood) is relatively low, but it can be much higher when the tidal movements are concentrated through narrow channels or around islands, headlands, or similar constraints. The enhanced streams allow power extraction using simple submerged turbines similar to those used for wind energy extraction. The prime advantage of tidal

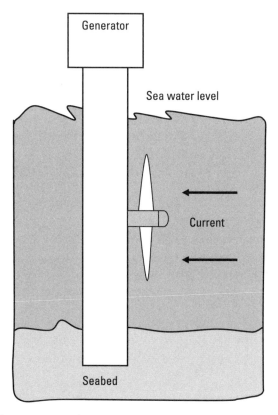

Figure 7.7 Tidal stream power plant.

stream over barrage systems is that they can generate power during both flood and ebb tidal periods [1].

Let us now derive an expression for the power delivered to a turbine by a water stream.

The kinetic energy in moving fluid (e.g., water or air) of mass and velocity is

$$E = \frac{1}{2}mv^2 \qquad (7.1)$$

Therefore, the power is

$$P = \frac{1}{2}\frac{m}{t}v^2 \qquad (7.2)$$

The mass m is the density ρ (kg/m³) times the volume V(m³), hence

$$P = \frac{1}{2}\frac{\rho V}{t}v^2 \qquad (7.3)$$

But the quantity $Q = V/t$ is the volume flow rate in cubic meters per second, therefore

$$P = \frac{1}{2}\rho Q v^2 \qquad (7.4)$$

However, Q is the product of the effective area of the fluid swept by the turbine times the speed of flow

$$Q = Av \qquad (7.5)$$

Therefore,

$$P = \frac{1}{2}\rho A v^3 \qquad (7.6)$$

Example 6.1

A tidal stream turbine has an efficiency of 20% and generates an output of 50 kW in a flow velocity of 2.5 (m/s). Calculate the diameter of the turbine. You may assume that the density of water is $\rho = 1000$ kg/m³.

Solution:

The output power is 50 (kW), hence the input power to the turbine is

$$P_{in} = \frac{P_{out}}{\eta} = \frac{50 \text{ kW}}{0.20} = 250 \text{ kW}$$

But the input power is given by

$$\frac{1}{2}\rho A v^3$$

Hence

$$250 \times 10^3 = \frac{1}{2}(1000)(2.5^3)A$$

and the area is

$$A = \frac{2 \times 250 \times 10^3}{1000(2.5^3)} = 32 \text{ m}^2$$

If the radius is R, and the area is πR^2, then

$$R = \sqrt{(A/\pi)}$$

The diameter is

$$2 \times \sqrt{(32/\pi)} = 6.38 \text{ m}$$

7.9 Calculations of Power Output of a Tidal Barrage

We will now derive a mathematical expression of the output power from a simple single-basin barrage system. To simplify the mathematics, we will assume that the basin is rectangular with cross-sectional area $A(\text{m}^2)$ and the high-to-low tidal range is $R(\text{m})$, as shown in Figure 7.8. The density of water is $\rho(\text{kg/m}^3)$ and the gravitational acceleration is $g(\text{m/s}^2)$.

Consider an infinitesimal strip of water whose height is dy located distance $y(\text{m})$ from the lower tide where $y = 0$. The mass of water (dm) contained in this strip is the volume of water times the density; that is

$$dm = \rho A dy \qquad (7.7)$$

The potential energy in this strip of water is

$$dE = ygdm \qquad (7.8)$$

or

$$dE = \rho A gy dy \qquad (7.9)$$

The maximum theoretical potential energy of the basin water per tide is therefore

$$E_{max} = \int_{y=0}^{y=R} \rho A gy dy = \frac{1}{2}\rho A gR^2 \qquad (7.10)$$

or

$$E_{max} = \frac{1}{2}mgR \qquad (7.11)$$

Now notice that the total mass of the water contained in the head R is

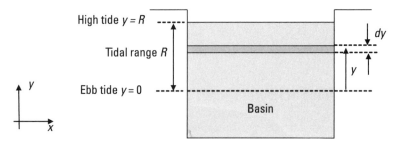

Figure 7.8   Calculation of a barrage output power.

$$m = \rho A R \tag{7.12}$$

so that the total potential energy you may have expected is $E = \rho A g R^2$. However, in reality only half of this energy is available. In the case of one-way turbine generation, this energy is available twice a lunar day (i.e., during a 24.8-hour period or $24.8 \times 60 \times 60 = 8.928 \times 10^4 \ s$). Hence, the maximum average power is

$$P_{max} = \frac{2\left(\dfrac{1}{2}\rho A g R^2\right)}{8.928 \times 10^4 \ s} = \frac{\rho A g R^2}{8.928 \times 10^4 \ s} \tag{7.13}$$

Assuming that the density of water is $\rho = 1000$ kg/m^3 and that the gravitational acceleration is $g = 10$ m/s^2, then

$$P_{max} = \frac{1000 \times 10 \ A R^2}{8.928 \times 10^4 \ s} = 0.1120 A R^2 \tag{7.14}$$

In the case of two-way turbines, this same energy is available four times in a day; hence, the maximum power is doubled

$$P_{max} = 0.224 A R^2 \tag{7.15}$$

Note the significance of the term R squared. This indicates that if the tidal range is doubled the amount of power is magnified by a factor of four. It must be stressed that the above calculation is for maximum theoretical average power. In practice, however, the actual power obtained will be about 25% of this theoretical maximum.

7.10 Advantages and Disadvantages of Tidal Power

The prime advantage of tidal power is that it is 100% predictable unlike solar or wind. In addition, tidal power is clean energy that uses free sustainable fuel (i.e., water). Barrages can help in certain sites in prevention of floods by using them to control high tides. Barrages could also help the local community by providing jobs, especially during the construction stage.

The main disadvantage of tidal plants is their impact on the marine wildlife. In addition, the construction of barrages across river estuaries can have negative effects on shipping and some consider them invasive.

7.11   Summary

Unlike most other renewable energy sources, tidal energy is not a solar form of energy. However, like other renewables, it is clean and sustainable because it uses water that is freely available and it does not affect the natural water cycle. In a tidal energy plant, the energy in water is harvested using barrages or lagoons and tidal streams. The prime advantage of tidal energy is that it is predictable unlike other renewables, such as wind and solar, because it is weather-independent. Tidal technology of barrages is well developed but barrages are expensive and can be invasive compared to tidal streams.

7.12   Problems

P7.1   Discuss two main advantages and disadvantages of tidal power systems.

P7.2   State the main advantage of tidal stream energy systems compared to barrage systems and discuss their main disadvantage compared to barrage systems.

P7.3   a. Using appropriate diagrams, explain the operation of a single basin tidal barrage power plant and explain why such power plants have difficulties in meeting typical variations in power demand during the course of a day.

b.   Explain how the above difficulty can be reduced using a double basin system.

P7.4   a. Explain the principle of operation of tidal stream systems for generating electricity.

b. Show that the power input to a tidal stream turbine is given by $P = \frac{1}{2}\rho A v^3$.

P7.5   Calculate the efficiency of each of the following tidal stream systems:

a. A turbine with a rotor of 18m in diameter, generating an output of 320 kW in a tidal flow of. Assume $\rho = 1.000 \times 10^3$ kg/m^3 and $g = 10$ m/s^2.

b. A power farm consisting of 5,000 small turbines, each with an area of 0.50 square meters, generating a total of 5 MW in a tidal flow of 2.5 m/s.

P7.6   Calculate the maximum energy stored in a single-basin tidal plant where the basin has an area of 20 km^2 at a location where the tidal range is 5.0m. Express your answer in MJ and in kWh. Assume $\rho = 1,000$ kg/m^3 and $g = 10$ m/s^2.

P7.7 A basin with a rectangle cross-sectional area, A, of 5 km by 10 km, has a two-way turbine. Estimate the maximum average power when the tidal range R is (a) 2.5m and (b) 3m. Assume $\rho = 1,000$ kg/m^3 and $g = 10$ m/s^2.

References

[1] Boyle, G. (ed.), *Renewable Energy Power for a Sustainable Future*, Oxford, UK: Oxford University Press with The Open University, 2012.

[2] Khan, M. J., G. Bhuyan, and M. T. Q. J. E. Iqbal, "Hydrokinetic Energy Conversion Systems and Assessment of Horizontal and Vertical Axis Turbines for River and Tidal Applications: A Technology Status Review," *Applied Energy*, Vol. 86, No. 10, 2009, pp. 1823–1835.

[3] National Ocean Service, U.S. Dept of Commerce, "Tidal Current," http://oceanservice.noaa.gov/education/tutorial_currents/02tidal1.html.

[4] World Energy Council, "Harnessing the Energy in Tides," *World Energy Council*, London, 2007.

8

Wind Energy

8.1 Learning Outcomes

After actively engaging with the material in this chapter, you should be able to

1. Explain the function of a wind turbine in converting the kinetic energy of a wind stream into rotational mechanical energy;

2. Define the terms cut-in, rated, and shutdown speeds, and the rated power of a wind turbine;

3. Derive the Betz limit on efficiency of a wind turbine;

4. Derive expressions for the input and output powers of a wind turbine;

5. Calculate the annual energy output from a given turbine using its power wind speed curve and the wind speed distribution for its location;

6. Discuss the main factors that determine the performance characteristics of a wind turbine;

7. Discuss the advantages and disadvantages of wind energy.

8.2 Overview

This chapter provides an introduction to the subject of wind energy. It presents a simple method for deriving mathematical expressions for the input and output powers of a wind turbine. This is followed by a discussion of vertical and horizontal wind turbines and their advantages and disadvantages. The Betz limit, which sets an upper bound on the theoretical efficiency of a wind energy conversion system, is also derived. Calculations of the annual energy output from a wind turbine using its power-speed curve and the distribution of the annual wind speed for its location is explained. Finally, several problems are presented at the end of the chapter to consolidate understanding of the material.

8.3   Introduction

A moving fluid, such as wind, has mass and velocity and therefore has kinetic energy that can be harvested and converted into a usable form of energy. In this chapter, we focus on the technology of harvesting wind energy and converting it into electrical energy, which is the most convenient form of energy. A wind turbine extracts the kinetic energy in a moving wind stream and converts it into rotational mechanical energy, which rotates an electric generator. This assembly of a wind turbine and a generator is a wind turbine-generator (WTG) set; however, we normally refer to it simply as a wind turbine.

Wind energy is another manifestation of solar energy. It is solar energy that heats different regions of the earth's surface to different temperatures, thus producing wind. The annual solar radiation received per unit area of the earth's surface in tropical regions is far more than that received by regions at higher latitudes. Due to the curvature of the earth, its surface becomes more slanted with increasing latitudes. Consequently, sunrays must travel longer distances to reach higher latitudes, thus losing more energy to the atmosphere before they reach the earth's surface. Variations in the temperature of the earth's surface between different regions results in variations in atmospheric pressure, which gives rise to the movement of air molecules, thus producing wind [1]. Wind energy is one of the most ancient forms of energy known to humans. It has been used to propel sailboats and grind grains for thousands of years. However, its use to generate electricity is relatively new; the first wind turbine to produce electricity was built in Scotland by James Blyth in 1887, and around the same time, a 12-kW fully functional turbine was built in the United States by Charles Brush [2].

8.4   Turbine Blades

The main part a wind turbine is the array of blades, typically two or three, fitted equally spaced around a shaft, as shown in Figure 8.1.

The blades are shaped so that the cross section of each blade, also included in Figure 8.1, is an airfoil all the way from the trailing to the leading edge. Because of this airfoil shape, a lift force is generated when wind blows over a blade,

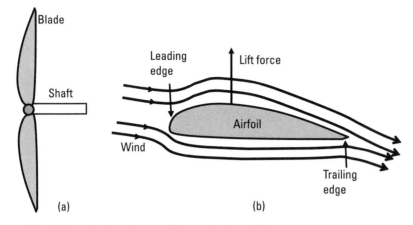

Figure 8.1 (a) Two-blade turbine, and (b) production of lift force.

which causes the blade to rotate. As the blades rotate, the kinetic energy in the wind is transferred to the turbine shaft (i.e., the rotor) as a rotational mechanical energy [3].

8.5 Types of Wind Turbines

One method of classifying wind turbines is in reference to the axis around which the turbine blades rotate. Consequently, a turbine is classified as horizontal axis wind turbine (HAWT) or vertical axis wind turbine (VAWT). In HAWTs, the axis of rotation is in line with the direction of wind, whereas in VAWTs the rotation axis is perpendicular to the wind direction. Wind turbines range in size from very small machines that produce few tens or hundreds of watts to very large turbines producing several megawatts. In general, in the context of wind electricity generation, unless otherwise stated, a wind turbine also includes an electric generator, a breaking system, a gearbox, and all the associated control devices.

8.6 Horizontal Axis Wind Turbines

HAWTs, which are used for electricity generation, typically have either two or three blades. The structure of a typical HAWT system is shown in Figure 8.2. The main components of the system are enclosed in the nacelle. The wind speed is normally much lower than the speed required to run the generator to generate electricity with the required frequency, 50 Hz in Europe and 60 Hz in the United States. Therefore, a gear box is used to step up the turbine speed before it is used to spin the generator. A wind turbine may also include an an-

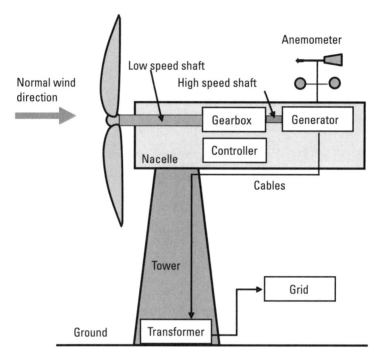

Figure 8.2   A typical HAWT and main components.

emometer, which measures the direction and speed of wind. The anemometer feeds the wind direction measurement as an electrical signal to the controller, which will send a signal to the yawing mechanism to rotate the nacelle so that the turbine will always be aligned to the wind direction for maximum efficiency of wind energy extraction.

This efficiency of wind energy conversion also depends on the angle of the wind-relative velocity known as the angle of attack, shown in Figure 8.3, which changes with the wind speed. Hence, a control signal representing wind speed is also sent to the controller to tilt the blades to keep them aligned for maximum energy extraction. Although not shown in the diagram, a wind turbine also includes a braking system for stopping the turbine in the event of an excessive wind speed that can damage the turbine if it is allowed to spin. The transformer is used to step up the output voltage of the generator to a level suitable for transmission and integration with the existing utility grid.

An important issue in the design of a horizontal wind turbine is whether to use an upwind or a downwind turbine, as shown in Figure 8.4. A downwind turbine has the advantage of letting the wind itself control the yaw, so the turbine automatically aligns itself correctly with respect to wind direction. However, there is a major problem with this design due to the wind shadowing effects of the tower. This is because every time a blade swings behind the tower,

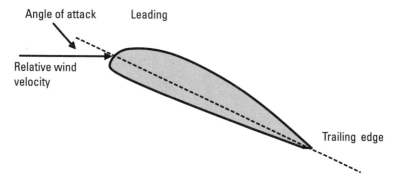

Figure 8.3   The angle of attack of the relative velocity.

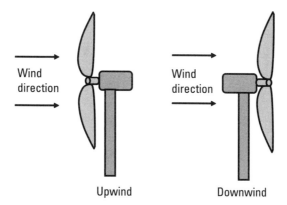

Figure 8.4   (a) Upwind and (b) downwind turbines.

it encounters a brief period of reduced wind. This causes the blade to flex, which increases the risk of blade failure, generates noise, and reduces the power output of the turbine. Upwind turbines, on the other hand, require complex yaw control systems to keep the blades facing the direction of the wind but are much quieter and more efficient. Currently, most electricity generating wind turbines in use are of the upwind type [4].

8.7   Vertical Axis Wind Turbines

Three different types of VAWTs are shown in Figure 8.5. The Savonius turbine was invented by the Finnish architect Laurentius Savonius in 1922. It consists of two semicylinders or cups. When the open cup is facing the direction of the wind, the force on it will be much larger than the force at the back of the other cup because the surface of the latter is rounded. This produces a dragging force

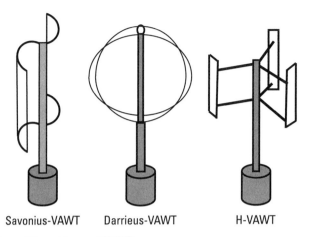

Savonius-VAWT   Darrieus-VAWT   H-VAWT

Figure 8.5 Three different types of VAWTs.

that will rotate the rotor. The Darrieus vertical axis turbine, named after its inventor, consists of a number of curved airfoil blades, thus it is a lift type of turbine. A Darrieus turbine runs much faster than a Savonius turbine; in fact, it can run at much higher speed than the speed of wind striking it, but the torque it develops is less than that of a Savonius turbine. The high-speed capability makes a Darrieus turbine suited for electricity generation. The H-type VAWT, also shown in Figure 8.5, has simple construction because it uses straight blades attached to a shaft using struts, which makes it easier to manufacture, and it is cheaper than the Darrieus turbine. In addition, the H-type turbine has better aerodynamic performance than the Darrieus type [5].

The main advantage of VAWTs is that they can harvest winds from any direction without the need for complex yaw control to continually reposition the rotor whenever the wind direction changes. The second advantage of VAWTs is that the heavy system components, such as the gearbox and generator, can be located on the ground, alleviating the need for structurally strong nacelle and strong tower as in the case with HAWT, which makes these components easily accessible for maintenance. The tower can be even made lighter and thus cheaper, particularly when guy wires are used. However, despite these advantages, they have found little commercial success to date due to several disadvantages [4]. The principle disadvantage of VAWTs is the fact that the blades are close to the ground. This is a major drawback because close-to-ground wind speeds are slower and since the power in the wind depends on the cube of the wind speed, this places considerable limitation on the output power of these turbines. Further, at excessively high wind speeds, when it becomes necessary to control the speed of a turbine to protect the generator, it is difficult to make the vertical blades spill the wind as is the case with pitch-controlled blades on the HAWT. In addition, wind is more turbulent the closer to the ground it is, which increases stress on the VAWT [6]. Another fundamental design decision for wind turbines relates to the number of rotating blades. Wind turbines with many blades operate with much lower rotational speed than those with fewer blades. As the revolution per minute (rpm) of the turbine increases, the turbulence caused by one blade affects the efficiency of the blade that follows. With fewer blades, the turbine can spin faster before this interference becomes excessive. A faster spinning shaft means that generators can be physically smaller in size for a given power rating. Most modern European wind turbines have three rotor blades, while American ones tend to have just two. Three-bladed turbines tend to be quieter; however, the third blade increases the weight and cost of the turbine by a considerable margin [4].

8.8 Power in the Wind

Since wind generates electricity, it is important that we can estimate the amount of power in a wind stream. The wind is a continuously moving fluid; hence, it is delivering energy continuously, and therefore we really need to determine the rate at which this energy is delivered by the wind (i.e., the power in watts). In order to evaluate the rate at which kinetic energy is delivered by wind, we need to know the mass of air reaching the turbine per second (i.e., the mass flow rate (m dot) $\dot{m}$(kg/s)). This requires the density (ρ) of air, which depends on temperature and altitude. Dry air at normal pressure at zero degrees Celsius has a density of 1.29 kg/m^{3}and at 15 degrees, its density is approximately 1.23 kg/m^3 [1]. Now let us determine how much energy is delivered by a wind stream to a turbine with radius R (m), as shown in Figure 8.6.

The kinetic energy of mass m (kg) of air traveling at speed of v(m/s) is

$$E_K = \frac{1}{2}mv^2 \qquad (8.1)$$

The area swept by the turbine is

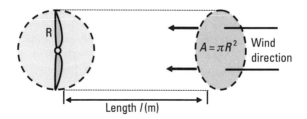

Figure 8.6 Determining the power delivered to a turbine.

$$A = \pi R^2 \tag{8.2}$$

Since the speed of wind is v(m/s), then in one second a cylinder of area A(m²) and length v(m) will pass the turbine. Hence, the volume of air passing the blades per second is

$$V = Av \quad (\text{m}^3/\text{s}) \tag{8.3}$$

and the mass of air passing the turbine per second (i.e., the mass flow rate $\dot{m}$ (kg/s)) is

$$\dot{m} = \rho Av \quad (\text{kg/s}) \tag{8.4}$$

Hence, the kinetic energy delivered by the wind to the turbine per second (in J/s or W) (i.e., the power) is

$$P_w = \frac{1}{2}\rho Av^3 \quad (\text{J/s}) \tag{8.5}$$

It is clear from the above relationship that the power in the wind is proportional to the density of the air, the area through which the wind is passing, and the cube of the wind velocity. Hence, velocity is very important for power extraction. It is imperative at this point to appreciate that the wind velocity, in the above expression for power, is the undisturbed wind velocity (i.e., it is the velocity that is unaffected by the turbine). However, not all of this power is converted to useful rotational mechanical energy by the turbine; as we will soon see it only represents the power in the wind.

8.8.1 Effect of the Number of Blades

Consider a point on the tip of a blade on the rotor of a turbine, as shown in Figure 8.7. As the blade rotates, its tip sweeps out a circle whose radius is R (m) (i.e., the length of the blade). This corresponds to an angle of 360 degrees or 2π radians. The distance traveled by the tip is the circumference of the circle (i.e., $2\pi R$). When the tip rotates through an angle of only one radian (i.e., approximately 57.3°), the tip has traveled through a distance of R (m) (see Figure 8.7). Similarly, when the rotor turns through one complete revolution, the tip travels a distance equal to $2\pi R$ (m). If the angular speed of the rotor is ω radians per second (rad/s), then the actual tip speed U in m/s is simply R times the angular velocity

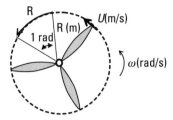

Figure 8.7 Angular and tip speeds.

$$U = \omega R \tag{8.6}$$

This is the tangential velocity of the rotor at the tip of the blades. Usually, the speed of a wind turbine is given in rpm. If the rotor is rotating at speed N (rpm), then it sweeps 2π radians every minute. Hence, its angular velocity in radians per second becomes

$$\omega = \frac{2\pi N}{60} \tag{8.7}$$

And the tip speed becomes

$$U = \frac{2\pi RN}{60} \tag{8.8}$$

By dividing the tip speed, U, by the undisturbed wind velocity, v(m/s) (i.e., the upstream of the rotor), we obtain a nondimensional ratio known as the tip speed ratio (TSR)

$$\text{TSR} = \frac{U}{v} \tag{8.9}$$

This ratio provides a useful measure against which the aerodynamic efficiency of a turbine can be assessed. The aerodynamic efficiency of a wind turbine is usually described by its power coefficient, which is defined as the ratio of mechanical power output from the turbine to the theoretical power in the wind. This quantity is given the symbol C_P and can be expressed as a decimal fraction or a percentage

$$C_p = \frac{\text{mechanical power output from the turbine}}{\text{theoretical power in the wind}} \qquad (8.10)$$

When the power coefficient is plotted against TSR for a number of turbines, such plots provide an effective way to present the performance of a given turbine and to compare wind turbines with differing characteristics [4].

A wind turbine of a particular design can operate over a range of TSR, but will usually operate with its maximum efficiency at a particular (i.e., optimum) TSR. This occurs when the velocity of its blade tips is a particular multiple of the wind velocity. This optimum TSR is commonly denoted as TSR_{max} with the corresponding power coefficient being C_{Pmax}. The optimum TSR for a given wind turbine rotor depends on both the number of blades and the width of each blade.

In order to extract energy as efficiently as possible, the blades have to interact with as much of the wind passing as possible through the rotor's swept area. The blades of a multiblade wind turbine interact with all the wind at very low TSRs, whereas the blades of a turbine with few blades have to travel much faster to virtually fill up the swept area in order to interact with all the wind passing through. If the TSR is too low, some of the wind travels through the rotor swept area without interacting with the blades, whereas if the tip speed ratio is too high, the turbine offers too much resistance to the wind, so that some of the wind goes around it. A two-bladed wind turbine rotor with each blade the same width as those of a three-bladed rotor will have an optimum TSR one-third higher than that of a three-bladed rotor. Optimum TSRs for wind turbines with a small number of blades range between about 6 and 20 [4].

In theory, the more blades a wind turbine rotor has, the more efficient it is. However, when there is a large number of blades in a rotor, the flow becomes more disturbed, so that they aerodynamically interfere with each other. Therefore, multiblade wind turbines tend to be less efficient overall than those with few blades. Three-bladed rotors tend to be the most energy efficient, two-bladed rotors are slightly less efficient, and one-bladed rotors slightly less efficient still. Wind turbines with more blades can be generally expected to generate less aerodynamic noise as they operate at lower tip speeds than wind turbines with fewer blades. The mechanical power that a wind turbine extracts from the wind is the product of its angular velocity and the torque imparted by the wind. For a given amount of power, the lower the angular velocity, the higher the torque, and conversely, the higher the angular velocity, the lower the torque. Conventional electrical generators run at speeds many times greater than most wind turbine rotors, so they generally require a gearing system. Wind turbines with few blades (e.g., two or three) are more suited to electricity generation because they operate at high TSRs, and thus do not require high gear ratios to match the speed of their rotors to that of the generator [4].

8.8.2 Maximum Theoretical Efficiency of Wind Turbines

Energy in the wind is a low-quality energy due to the randomness of its distribution. Whenever we convert a low-quality energy into a high-quality energy, such as converting wind energy to a rotational mechanical energy, there will always be a fundamental physical limit on the maximum theoretical efficiency of such a conversion. In the wind-to-mechanical energy conversion, the limit has been derived by German physicist Albert Betz in 1919 and is known by his name as the Betz limit. Let us now derive this limit using a simplified approach. As shown in Figure 8.8, wind approaches (upwind) the turbine with its undisturbed speed v(m/s). This is the same speed at some distance far away from the turbine (i.e., unaffected by the presence of the blades). As the wind hits the blades its speed is reduced by the blades to v_b(m/s) because the blades have extracted some of its kinetic energy. The reduction of the wind speed results in a reduction of its pressure; hence, it expands resulting in the tapered stream tube. As the wind leaves the blades (i.e., the downwind) its speed is v_d (m/s). If the downwind speed is the same as the upwind speed (i.e., if $v = v_d$) then the turbine has not extracted any energy from the wind. If, on the other hand, the turbine extracted all the energy in the wind then the wind speed will be zero at the turbine and therefore no more wind can be allowed to flow (i.e., the power extraction by the turbine will stop). Hence, an ideal turbine must slow the undisturbed wind speed by an optimum amount. In fact, Betz showed that an ideal wind turbine would reduce the wind speed to one-third of its original undisturbed value.

The power extracted by the turbine blades is given by the difference in rate of kinetic energy between the upwind and downwind flows

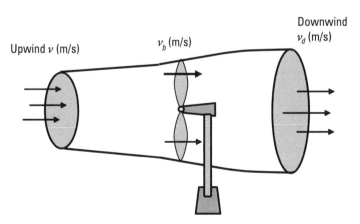

Figure 8.8 Wind approaching the turbine with undisturbed wind speed.

$$P_b = \frac{1}{2}\dot{m}(v^2 - v_d^2) \tag{8.11}$$

This assume that the mass flow rate $\dot{m}$ within the tube is the same everywhere. The mass flow rate at the plane of the rotor, where the swept cross-sectional area is the area of the rotor, is

$$\dot{m} = \rho A v_b \tag{8.12}$$

and

$$P_b = \frac{1}{2}\rho A v_b (v^2 - v_d^2) \tag{8.13}$$

Now we make the assumption that the wind speed at the rotor is the average of the upwind and downwind speeds

$$v_b = \frac{1}{2}(v + v_d) \tag{8.14}$$

Therefore, we can write

$$P_b = \frac{1}{2}\rho A \frac{(v + v_d)}{2}(v^2 - v_d^2) \tag{8.15}$$

Let us define the ratio of the downwind speed to the upwind speed as

$$\lambda = \frac{v_d}{v} \tag{8.16}$$

Hence

$$P_b = \frac{1}{2}\rho A(\frac{v + \lambda v}{2})(v^2 - \lambda^2 v^2) \tag{8.17}$$

Therefore

$$P_b = \frac{1}{2}\rho A v^3 \left[\frac{1}{2}(1+\lambda)(1-\lambda^2) \right] \tag{8.18}$$

The quantity in the square brackets represents the fraction of the power extracted from the wind. This is the power coefficient and is defined as

$$C_p = \left[\frac{1}{2}(1+\lambda)(1-\lambda^2) \right] \tag{8.19}$$

That is, the power available for extraction by the turbine is

$$P_b = \frac{1}{2}\rho A v^3 C_p \tag{8.20}$$

A plot of the variations of the power coefficient C_P with λ is shown in Figure 8.9, which reveals that the power coefficient is maximum when λ is 1/3. This can be verified analytically as follows.

To find when this fraction is maximum, we differentiate with respect to λ:

$$\begin{aligned}
\frac{dC_p}{d\lambda} &= \frac{1}{2}\left[(1+\lambda)(-2\lambda) + (1-\lambda^2) \right] \\
&= \frac{1}{2}\left[-3\lambda^2 - 2\lambda + 1 \right]
\end{aligned} \tag{8.21}$$

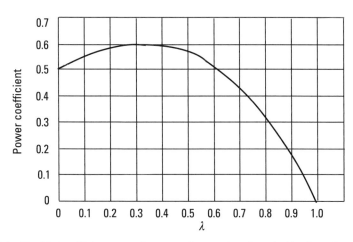

Figure 8.9 The blade efficiency reaches a maximum when the wind is slowed to one-third of its upstream value.

Now we equate this derivative to zero and solve for λ

$$\frac{1}{2}\left[-3\lambda^2 - 2\lambda + 1\right] = 0 \quad or \quad \left[-3\lambda^2 - 2\lambda + 1\right] = 0 \qquad (8.22)$$

$$\therefore (1+\lambda)(1-3\lambda) = 0 \quad \therefore \lambda = -1, \ and \ \lambda = \frac{1}{3} \qquad (8.23)$$

Therefore, the solution is

$$\lambda = \frac{1}{3} \quad i.e.\ v_d = \frac{v}{3} \qquad (8.24)$$

That is, the blade efficiency is maximum if it slows the wind speed to one-third of its undisturbed upstream velocity. Now substitute this value of $\lambda = 1/3$ in the equation for the power coefficient (also known as the rotor efficiency)

$$C_p = \left[\frac{1}{2}(1+\lambda)(1-\lambda^2)\right] = \left[\frac{1}{2}(1+\frac{1}{3})(1-\frac{1}{9})\right] = \frac{16}{27} = 0.593 \quad or \quad 59.3\% \qquad (8.25)$$

Therefore, we conclude that the maximum theoretical efficiency of a rotor is 59.3%. This is called the Betz efficiency or Betz law. However, wind turbines cannot operate at this theoretical maximum limit, and currently common rotor efficiencies are in the region of 35% to 45%. For a given rotor speed the efficiency of the rotor depends on the speed of the rotor. If the rotor turns very slowly, then efficiency is reduced since the blades are allowing too much wind to pass without extracting any power from it. Similarly, if the rotor runs too fast then the efficiency is again reduced because of the turbulence caused by one blade adversely affects the efficiency of the next one in power extraction. In addition, by the time we consider the other factors in a complete wind turbine system (e.g., the gearbox, bearings, generator, and so on) only 10%–30% of the power of the wind is ever actually converted into usable electricity.

8.9   Electrical Power and Energy from Wind Turbines

The output power of a wind turbine varies, among other things, with wind speed, and every wind turbine has its own characteristic power-speed curve, which is provided in the turbine manufacturer's datasheet. A typical power-speed curve is shown in Figure 8.10. The shape of a power-speed curve, or simply the power curve, is influenced by several factors, such as rotor swept area, choice of airfoil, number and shape of blades, and the optimum tip-speed ratio. We define the cut-in speed as the speed after which the turbine starts to generate power. Note that although a turbine may start rotating at slightly lower speed than the cut-in, the amount of power it generates may be just enough to supply its losses; hence, there is no point of starting it to simply feed losses. As the speed of the turbine blades increases above the cut-in speed, the output power is proportional to the cube of the speed, and then rises sharply. This continues until the wind speed reaches the rated speed of the turbine and the turbine starts to supply its rated power. If the wind speed is increased any further, the output power is limited to the rated power to prevent damage to the turbine and generator. If the wind speed increases to the shutdown speed, the turbine is automatically stopped by its built-in breaking mechanism.

The energy that a wind turbine will produce depends on both its wind power-speed curve and the wind speed frequency distribution at the site where it is located. The latter is essentially a graph or histogram showing the number of hours in which the wind blows at different wind speeds during a given period. Figure 8.11 shows a typical wind speed frequency distribution. For each incremental wind speed within the operating range of the turbine, the energy produced at that wind speed can be obtained by multiplying the number of hours of its duration by the corresponding turbine power at this wind speed

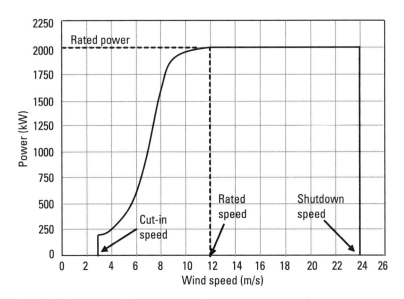

Figure 8.10 Typical wind turbine power-speed curve.

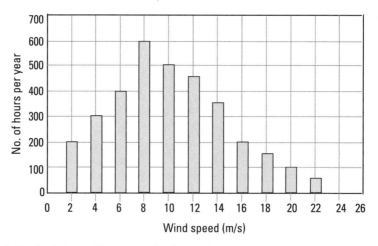

Figure 8.11   A wind speed frequency distribution for a typical site.

given by the turbine's power-speed curve. This data can then be used to plot a wind energy distribution, such as that shown in Figure 8.12.

The total energy produced in a given period is then calculated by summing the energy produced at all the wind speeds within the operating range of the turbine.

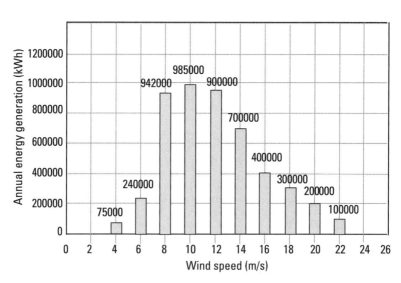

Figure 8.12   Wind energy distribution for the same site as in Figure 8.11, showing energy produced at this site by a wind turbine with the power curve in Figure 8.10.

8.10   Advantages and Disadvantages of Wind Energy

The principle advantage of wind energy is that it uses free fuel (i.e., wind), which is unlimited, and harvesting its kinetic energy does not affect the natural cycle of wind. It is clean and does not produce any greenhouse gases. Disadvantages of wind energy include the relatively high initial costs, particularly for offshore plants. Also, wind turbines can upset wildlife and can be dangerous to birds. Generally, wind turbines must be installed in locations far away from places where the generated energy is needed. This adds to the cost of power transmission and increases losses, which consequently reduces the overall system efficiency. In addition, it increases the cost of maintenance and servicing, specifically for offshore turbine. Finally, wind turbines generate audio and electromagnetic noise, which may not be acceptable to people living nearby and may also harm wildlife. Hence, more research is needed to improve the performance and reduce the cost of wind energy.

8.11   Wind Turbine Generators

Conventional power plants use a synchronous generator, which runs at a fixed speed. Wind energy plants, however, use an induction generator, which is very robust and has simple construction and therefore is more reliable [7]. An induction machine does not require any slip rings or brushes and thus, requires little maintenance compared to the synchronous machine. For wind energy systems, the principle advantage is that an induction machine can run at variable speed as a motor and as a generator. In wind energy systems, an induction machine is used in both modes; as a motor to start spinning the turbine blades and as speed picks up it is switched to generator mode.

8.12   Summary

Wind energy is a renewable source of energy; it is clean and uses wind as fuel, which is sustainable and free. Kinetic energy in a wind stream is harvested and converted to rotational mechanical energy that turns an electric generator using a wind turbine energy conversion system. In this system, there is a theoretical upper bound on the efficiency of the rotor, which is approximately 59%. However, taking losses in various components of the conversion system into account, the overall efficiency is between 10% to 30%. There are two main types of wind turbines: horizontal and vertical axis turbines. HAWTs are the predominant type used in generating electricity. Wind energy conversion systems use induction generators mainly because they are variable speed machines and can run as motors as well as generators.

8.13   Problems

P8.1   a. Write a concise explanation of the statement "wind energy is a form of solar energy".

b. Discuss the advantages and disadvantages of vertical and horizontal wind turbines.

c. Discuss the significance of the blade length and the number of blades in a HAWT paying attention to the output power and application of the turbine.

d. Explain the significance of the power coefficient of a wind turbine.

e. Explain the difference between a drag and a lift type wind turbine.

f. Sketch a typical power-speed curve of a wind turbine and use it to define the following terms: cut-in speed, rated speed, shutdown speed, and rated power of a wind turbine.

g. Define the tip speed and TPR.

h. Discuss two advantages and two disadvantages of wind energy.

i. Explain the dependence of the efficiency of a wind turbine rotor on its speed.

P8.2   Making any reasonable assumptions, derive from first principles an expression for the power delivered to a wind turbine from the kinetic energy of wind.

P8.3   Making any reasonable assumptions, show that the upper bound on the efficiency of a wind turbine rotor is 16/27.

P8.4   A HAWT has a blade length $l = 35m$ and power coefficient C_p = 0.4. Determine the available rotational power from the turbine when the wind speed is 12m per second and the air density is 1.25 kg per cubic meter.

P8.5   A wind turbine has a diameter of 50 meters. If the wind speed is 12 m/s and the air density is 1.28 kg per cubic meter, how much power does the wind deliver to this turbine?

P8.6   A wind turbine generates an output power of 120 kW when the wind speed is 10 m/s. The turbine diameter is 30m, and the air density is 1.24 kg per cubic meter. Determine the power delivered by the wind and the efficiency of the system.

P8.7 The power-speed curve for a turbine is shown in Figure 8.13.

a. What is the value of each of the following: the rated power of the turbine, the cut-in speed, and the shutdown speed?

b. What is the output power at speed of 6 and 12m per second? Comment on the answers.

c. What is the output power at speed of 8 and 16m per second? Comment on the answers.

P8.8 A HAWT generates 220 kW when the air density is 1.25 kg per cubic meter and the wind speed is 12 m/s. The length of each blade is 15m.

a. Determine the tip speed and the speed of the turbine in revolutions per minute if the tip speed ratio is 4;

b. Determine the required gear ratio if the generator needs to run at 1500 rpm;

c. Determine the efficiency of the overall system.

P8.9 The turbine whose power-speed curve given in Figure 8.13 is located at a site where the annual wind speed distribution is shown in Figure 8.14. Determine the annual electrical energy generated by the turbine.

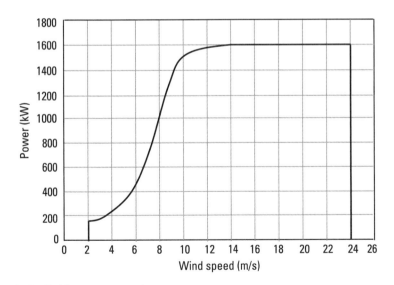

Figure 8.13 Turbine power-speed curve.

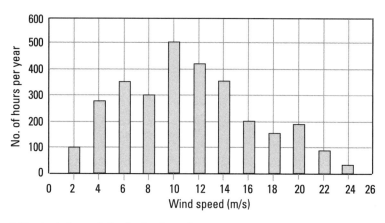

Figure 8.14   A wind speed distribution for a site.

References

[1]   Boyle, G. (Ed.), *Renewable Energy Power for a Sustainable Future*, Oxford, UK: Oxford University Press with The Open University, 2012.

[2]   Price, J. T., "James Blyth Britain's First Modern Wind Power Pioneer," *Wind Engineering*, Vol. 29, No. 3, 2005, pp. 191–200.

[3]   Chen. J., and Q. Wang, *Wind Turbine Airfoils and Blades: Optimization Design Theory*, Beijing: De Gruyter, 2018.

[4]   Masters, G. M., *Renewable and Efficient Electric Power Systems*, Hoboken, NJ: John Wiley & Sons, Inc., 2004.

[5]   Ferreira, C. S., A. H. Madsen, M. Barone, B. Roscher, P. Deglaire, and I. Arduin, "Comparison of Aerodynamic Models for Vertical Axis Wind Turbines," *Journal of Physics*, Vol. 524, No. 1, 2014.

[6]   Apelfröjd, S., S. Eriksson, and H. Bernhoff, "A Review of Research on Large Scale Modern Vertical Axis Wind Turbines at Uppsala University," *Energies*, Vol. 9, No. 7, 2016, p. 570.

[7]   Hughes, E., and J. Hiley, *Hughes Electrical and Electronic Technology*, Harlow, UK: Pearson Education Limited, 2016.

9

Bioenergy

9.1 Learning Outcomes

After actively engaging with the material in this chapter, you should be able to

1. Discuss the main types of biomass fuels;
2. Explain how biomass resources store energy from the sun;
3. Explain with examples the meaning of primary and secondary biomass fuels;
4. Define the terms energy density, biomass yield, and carbon-neutral energy;
5. Calculate the yield of a biomass crop from a given area;
6. Calculate the energy content of a given quantity of biomass fuel;
7. Discuss advantages and disadvantages of biomass as a source of sustainable energy;
8. Explain the need for processing biomass materials;
9. Calculate the cost per unit of electricity from a biomass power plant;
10. Use available literature to critically contribute to the debate around using land for energy crops instead of food.

9.2   Overview

This chapter presents an introduction to bioenergy as a source of renewable energy. It defines and explains important terms and concepts relating to bioenergy and biomass fuels. A discussion of primary and secondary biomass materials and their use as feedstock for bioenergy production is presented. Important parameters relevant to the production of bioenergy, such as the yield and energy density of biomass materials, are explained with numerical examples. Advantages and disadvantages of bioenergy are also discussed and the processes needed to convert biomass materials into suitable forms for energy production are outlined. Finally, several problems to check and consolidate understanding of the material are provided at the end of the chapter.

9.3   Introduction

Carbon-rich substances, such as sugarcane, maize, and straw are good stores of chemical energy; hence, such substances are also known as carbon-based fuels. Energy stored in these substances can be released during combustion (i.e., burning). During combustion, these substances take in oxygen (O_2) from the atmosphere and release carbon dioxide (CO_2) and heat energy. Carbon-based fuels are of two types: fossil fuels and biofuels. Fossil fuels such as coal and oil are derived from carbon-rich matter that has taken millions of years to form, while biofuels are derived from organic matter that has been recently grown or formed and are known as biomass. Biomass is the name given for all living, or recently living, matter originating from plants and animals that represent a source of energy. The term bioenergy is used to refer to fuels derived from biomass. It includes solid fuels such as wood, liquid fuels such as biodiesel, and biogases such as bioethanol. Bioenergy is currently used in transport, heating, cooking, and for biopower (i.e., generating electricity). Examples of biomass include forestry products such as wood, agricultural by-products such as straw, industrial waste such as woodchips and wood off-cuts, animals, and humans waste, and plants grown specifically to be used as biofuels such as miscanthus, sugarcane, and maize. Traditional biomass (e.g., the burning of wood) is the oldest source of energy known to humans and it has been used since humans learned how to start a fire. In fact, traditional biomass is still a significant source of energy in many developing countries where wood is still used for heating, cooking, and lighting. For example, in some remote areas of some developing countries, horse manure is still being collected, dried, and used as fuel in special clay ovens for cooking. This is in addition to using it as a fertilizer.

    Bioenergy is the fourth largest source of energy after oil, natural gas, and coal and it accounted for approximately 10% of the total global demand of primary energy in 2015 [1]. Moreover, bioenergy represented the largest share of about 70% among all renewable energy sources providing about 57 EJ. This is about 10% of the world's total primary energy supply in 2016 [2]. In addition, biomass feedstock was the third largest renewable energy source for electricity generation in 2016 [3].

Biomass represents all living things on Earth that is contained in a thin surface layer of the earth known as the biosphere. The mass of the bioshpere is a very tiny fraction of the mass of the earth, yet it represents a huge store of chemical energy that is received from the sun, and much of it is available for us to harvest. Estimating the global potential of biomass energy and utilization is a very complex task due to the numerous forms, locations, and types of biomass feedstock [4]. However, some research estimates that the rate of energy storage by the total land biomass is about 2400 EJ per year, which is about four times the current global demand of primary energy [5]. Another study consisting of 19 different assessments concluded that energy crops alone should be able to provide about 1,300 EJ per year in 2050 [6].

It is important, however, to appreciate that although the biosphere represents a tiny fraction of the mass of the earth, its existence is so vital for life on this planet. If all plants were to disappear from the face of the earth, there would be no gases, such as oxygen and nitrogen, which are essential to sustain life.

9.4 Biomass as a Renewable Energy Source

A source of energy can only be considered renewable if it is sustainable or can be replenished in relatively short periods of time and its use does not lead to any significant emissions of greenhouse gases. Well, as we will soon discover, biomass fits, or at least it can be made to fit, this definition. Biomass is almost a carbon-neutral source of energy (i.e., the net release of carbon dioxide in the atmosphere) and is negligible when used responsibly for energy production. This is best illustrated with a comparison between the methods of extracting energy from a fossil fuel such as coal, and from a biomass source, such as a tree. Millions of years ago plants such as trees lived on the surface of Earth and while living, they absorbed carbon dioxide from the atmosphere and energy from the sun for several years. When they died, decayed, and compressed deep underground, they stored some of this energy and carbon dioxide and turned into coal. When coal is burned, this energy and carbon dioxide that has been stored for millions of years is released. Clearly, since it took millions of years to form coal, and therefore it is not a sustainable source of energy. In fact, coal is expected to be depleted during the next 30 to 80 years along with natural gas and oil [7]. On the other hand, if we plant a tree and leave it for several years to grow, say 5 to 10 years, not millions, during this time it will be absorbing carbon dioxide from the atmosphere and energy from the sun and storing them. When the same tree is cut, it is processed to convert it into suitable form for use as a source of energy (e.g., wood logs or wood pellets). Later, when the tree is burned, it can only release the carbon dioxide it absorbed from the atmosphere while it was alive along with the stored energy. But if we plant another tree, the new tree will absorb a similar amount of carbon dioxide as that released by burning the first tree. Hence, if we plant a new tree for every tree we burn, the process should be carbon-neutral. Otherwise (i.e. if we burn more trees than we plant) this carbon neutrality will not hold and more carbon dioxide will be released than absorbed, which will lead to deforestation. However, carbon dioxide will also be released during processing and transporting of trees, and therefore, we must plant more trees than we use for energy in order to offset this extra carbon dioxide emission. Hence, fundamental to keeping this source of energy sustainable and carbon-neutral is a well-managed strategy for plantation and cultivation of trees.

9.5　Biomass as a Solar Energy Source

Biomass is a manifestation of solar energy that is stored in carbon containing compounds in biomass materials. Plants store energy from the sun during the process of photosynthesis. During this process, plants take in carbon dioxide and water from the atmosphere and energy from the sun, which is what makes plants biomass sources of fuel. During the process of photosynthesis, six carbon dioxide molecules react with six water molecules and with input energy from the sun according to the chemical equation

$$6\,CO_2 + 6\,H_2O + \text{light energy} \; \rightarrow \; C_6H_{12}O_6 + 6\,O_2 \tag{9.1}$$

The results of this reaction are glucose ($C_6H_{12}O_6$), which is a carbohydrate, and oxygen, which is released into the atmosphere. This is a constant cycle during which a plant absorbs water, carbon dioxide, and solar energy from the surroundings and releases oxygen into the atmosphere. This cycle is broken when a plant is burned, taking in oxygen and releasing carbon dioxide and heat energy. It is important to appreciate that carbon is a key element in biomass and is present in many compounds that make up the bulk of all living tissues. Examples of carbon-rich compounds found in biomass include glucose, cellulose, hemicellulose, lignin, lipids, and proteins. Compounds that contain only carbon, hydrogen, and oxygen are collectively known as carbohydrates and are the prime stores of energy in biomass.

9.6 Biomass as a Fuel

Extraction of energy from biomass normally requires combustion, which involves taking in oxygen and releasing heat energy and carbon dioxide. As an example, consider the combustion of methane (CH_4), which can be produced from biomass. A methane molecule consists of one carbon atom and four hydrogen atoms and each oxygen molecule consists of two oxygen atoms (O_2). During combustion reaction, each methane molecule reacts with two oxygen molecules according to the chemical equation

$$CH_4 + 2O_2 \rightarrow CO_2 + 2H_2O + \text{Energy} \qquad (9.2)$$

On the input (left-hand) side of the equation energy is stored as chemical energy in the fuel (i.e., in the methane) and oxygen, and on the output side (the right hand) of the equation, energy is released in the form of heat energy plus chemical energy in the carbon dioxide and water. The difference between the input and output energies define the energy or heat content, also known as the energy value of the fuel, in this case methane. When burning a fuel, if we know the chemical formula of the fuel and the relative atomic masses of the chemical elements of its molecules, we can calculate the amount of carbon dioxide emitted when a given amount of fuel is burned. For example, during the combustion of methane (CH_4), the relative atomic masses of the atoms of carbon and oxygen are 12 times and 16 times the mass of a hydrogen atom, respectively. This can be represented in the balanced chemical equation as

$$
\begin{array}{ccccccc}
CH_4 & + & 2O_2 & \rightarrow & CO_2 & + & 2H_2O \\
12 + (4 \times 1) & + & 2 \times (2 \times 16) & \rightarrow & 12 + (2 \times 16) & + & 2 \times (2 \times 1 + 16)
\end{array}
\qquad (9.3)
$$

That is, burning 16 tonnes of methane releases 44 tonnes of carbon dioxide. In other words, burning one tonne of methane produces 2.75 (=44/16) tonnes of carbon dioxide. The difference in energy between the input and output sides of the equation is about 55 GJ per tonne. Hence, we say that the energy content of methane is 55 GJ per tonne or 55 MJ/kg. In comparison, coal has an energy content of 24 GJ/tonne and is one of the highest producers of carbon dioxide because its chemical composition is largely carbon. Another common name for the energy content of fuel is the energy density, which we can now define as the amount of energy released when burning one kilogram of a given substance and is usually expressed in MJ/kg or in kWh/kg. The energy density for some fuels is provided in Table 9.1 [8, 9].

Table 9.1

Energy Density of Some Fuels

Fuel	Energy Density by Mass (GJ/t)	Energy Density by Volume (GJ/m³)
Wood (oven-dried)	18	9
Coal (domestic)	28	25
Miscanthus	13	2
Maize grain (air-dried)	19	14
Wood (air-dried)	15	9
Domestic waste	9	1.5
Wood chips (30% MC)	12.5	3.1
Domestic heating oil	43	36
Baled straw	15	1.5
Sugarcane residue	17	10
Paper (stacked newspaper)	17	9

Example 9.1

Calculate the volume of airdried wood that is needed to bring one liter of pure water from 25°C to the boiling point (i.e., to 100°C). Given that the specific heat of water = 4,200 J /kg K, the mass of 1 liter of water is 1 kg, the energy density of airdried wood is 15 MJ/ kg, and the density of airdried wood is 600 kg/ m^{-3}.

Solution:

Heat energy needed = mass × temperature rise × specific heat capacity

Hence,

$$\text{heat energy needed} = 1 \text{ kg} \times (100\text{-}25) \text{ °C} \times 4200 \text{ J/kgC}$$
$$=315 \text{ kJ}$$

The energy released by burning 1 kg of wood is 15 MJ, but we only need 315 kJ, so the required mass of wood is $\dfrac{315 \text{ kJ}}{15000 \text{ kJ/kg}}=0.021 \text{ kg}$

$$\text{Volume}=\frac{\text{mass}}{\text{density}}$$

and the volume of the required wood is

$$V = \frac{0.021\,\text{kg}}{600\,\text{kg/m}^3} = 3.5 \times 10^{-5}\,\text{m}^3 \ \text{ or } 35\,\text{cm}^3$$

Example 9.2

Repeat Example 9.1 above using miscanthus instead of airdried wood, given that its energy density by volume is 2 GJ/m³ and the density of water is 1,000 kg/m³ and explain your answer. Note that 1 liter of water is 0.001 cubic meter.

9.7   Primary Biomass Energy Sources

In this section, we will briefly consider various primary biomass sources of energy. Primary sources refer to plants that have been specifically grown to be used as biofuels. This means that a primary source requires a site that has been set aside for growing the biomass source. We will consider woody biomass derived from woody plants, cellulosic, sugary, and oilseeds crops in addition to microorganisms such as microalgae, which are a class of their own [8]. The key element in all of these is carbon, which represents the energy store.

9.7.1   Woody Biomass

The main source of woody biomass are trees that consist of mainly carbohydrates, lignin, cellulose, and hemicellulose. Woody biomass come from trees that have been planted specifically for energy production in specially designated forests. Well-managed planation and harvesting of trees for energy can provide a sustainable fuel source with reduced carbon dioxide emissions into the atmosphere.

Example 9.3

Use data in Table 9.1 to answer this question. The volume of one tonne of a certain grade of coal is 0.53 m³. Calculate the volume of miscanthus that would be required to supply the same amount of energy as one tonne of coal.

Solution:

From Table 9.1, the energy content per tonne of coal is 28 GJ/t and the energy content per m³ of miscanthus is 2 GJ/m³; hence, to obtain 28 GJ from miscanthus we need

$$28 \div 2 = 14\text{m}^3$$

Note that 14 cubic meters of miscanthus are required in order to match the energy provided by just over half a cubic meter of coal. This low energy density of biomass can be a problem as it means higher costs of both storage and transport.

Example 9.4

Repeat Example 9.3 above for airdried wood instead of miscanthus. Explain your answer.

9.7.2   Cellulosic Materials

Cellulosic bioenergy crop is a main constituent of straw and many other major food crops. The most commonly known cellulose crop is miscanthus, also known as elephant grass. Miscanthus is flowery plant that originated in Asia and Africa and is currently grown in Europe and the United States. It has a low mineral content and high biomass yield and can be harvested annually, which makes it a favorite choice as a biofuel, outperforming maize and many other alternatives. The annual yield of miscanthus can be as much as 18 tonne per hectare with an energy density of 17 MJ/kg. However, these figures can vary considerably depending on factors such as location, climate, and nature of soil.

9.7.3   Sugary Crops

Sugar crops such as sugar beet, maize, and sugarcane can be grown to provide feedstock for energy production. The main biomass energy constituent of these is either sugar or starch, which when fermented produces bioethanol fuel. Sugarcane grown specifically for energy can be genetically modified to increase its fiber content. This results in significant increase in its biomass yield [10]. The high biomass yield of sugarcane and its high bioethanol productivity are major factors that contribute to the balance of greenhouse gas emission [11]. The residue of processing sugar crops for energy can be used as animal feed, which is an added advantage. In some locations, the yield of sugarcane can be as much as 98 tonne per hectare with an energy content of 16 MJ/kg [12]. Sugarcane is a significant crop for both energy and food, and it is estimated that 2.2 million hectares worldwide are used for its production. In Brazil for example, sugarcane annual business amounts to more than $20 billion [12].

9.7.4   Oilseed Crops

Oilseed crops such as soya beans, rapeseed, and sunflower seeds are grown primarily for the oil in their seeds, which has high levels of energy-rich fats as well as proteins. They have been grown for over thousands of years as food products. More recently, interest has been growing in these oily crops not only due to their health benefits, but also for their use as feedstock to produce bioenergy. It is estimated that the global production, for example, of vegetable oils in 2006 was around 110 million tonnes [13]. Moreover, the oil from all these crops has many uses in industries such as cosmetics. The extraction of the oil from the seeds of these crops is by pressing, which a simple and inexpensive technique. The main bioenergy that is currently obtained from these oils is biodiesel, and the residue left from the oil extraction process can be used as animal feed. Typical yield of oily crops is approximately 8 to 15 tonnes per hectare per year [8].

9.7.5 Microalgae

Microalgae, found naturally in fresh waters that are currently farmed for energy production, have a high level of energy-rich fats and thus can be used as a bioenergy source, for example to produce biodiesel and biopetroleum. There are several attractive features that make microalgae good source of sustainable energy, such as the fact that they grow in water; hence, they don't have to compete for land that may otherwise be used for food production. There are also claims that they clean water of pollutants and help absorb carbon dioxide [8].

Grossly estimated yields of different annual biofuels per hectare are given in Table 9.2 [8]. It must be noted that these figures are only for guidance as actual figures depend on many factors such as the moisture content (MC), location, nature of soil, and climate. In fact, MC is a very important factor in determining the yield of a biomass and it is the norm that the yield is quoted in terms of dry tonnes with 0% moisture (i.e., oven-dried biomass).

9.8 Secondary Biomass Sources

A secondary biomass source refers to a source resulting from nonenergy use of biomass but has the potential for energy production. In other words, this is a biomass source that comes from wastes or refuse, such as industrial wastes, agricultural wastes, by-products, household refuse, and unwanted products of human activities. Examples of secondary sources include straw, which is an ag-

Table 9.2

Typical Approximate Yield of Dry Matter in Tonne Per Hectare Per Year

Fuel category	Examples	Yield of Dry Matter (t/ha.yr)
Woody biomass	Trees, hardwood	10 to 20
Cellulosic biomass	Miscanthus, seaweeds	10 to 60
Starch/sugary crops	Maize, sugarcane	10 to 35
Oily crops	Oilseeds (e.g., rape, sunflower)	8 to 15

ricultural by-product, wood off-cuts, rice husk, and human and animal wastes. Note that industrial waste, such as plastics, that have been synthesized from fossil fuels are not considered sources of renewable energy [8].

9.8.1   Wood Residues

It is estimated that the timber industry leaves behind about 15% of tree crops as wood waste on site, which can be utilized for energy production. Such wood residue may be used as firewood for power and heat or for making papers. In some countries like Sweden, these residues have been used for many years at an annual increasing rate of 10% [8].

9.8.2   Crop Coproducts

It is estimated that the annual global energy content of the residue form wheat and maize, the two major cereal crops, amount to 15 to 20 EJ. Similarly, the total energy content of the annual residues of sugarcane and rice is estimated at about 18 EJ. Increased reclamation of wastes and increased efficiency of conversion to electrical energy could result in up to 50 GW of generating capacity from the global sugar industry alone [8].

9.8.3   Animal Wastes

Cattle deposit their manure in the fields where they graze, which later decomposes aerobically through the respiration of a range of organisms that feed on it. This process takes in oxygen and releases carbon dioxide in the atmosphere, which is essentially a carbon-neutral process. However, wet slurry that is stored in bulk decomposes anaerobically and consequently releases methane and nitrous oxide, which can be dangerous to humans and the environment. However, under well-managed and controlled anaerobic digestion, biogas can be generated for energy use. Another common animal waste is poultry litter, which has relatively low moisture content allowing direct combustion with an energy content of 9 to 15 GJ per tonne [8].

9.8.4   Municipal Solid Waste

It is estimated that the average energy content of household waste, which is collected as municipal solid waste (MSW), is about 9 GJ per tonne. This everyday waste, which includes items such as packaging, newspapers, unwanted food, and clothing, represents a significant source of energy, while simultaneously helps keep the environment clean and reduces the emission of harmful gases into the atmosphere. In the industrialized countries, the average household generates about 1 tonne of MSW per year and in the United Kingdom about

20 million tonnes of food waste is generated each year [8]. In 2013, American households generated 254 million tonnes of MSW and recycled about 30% of it [14]. The United Kingdom recycled about 46% of the MSW generated in 2017 [15]. It is estimated that between 30%–60% of MSW is recycled in most western European countries and it is estimated that 24 waste-to-energy plants operating in England were treating 4 million tonnes of MSW in 2012 [16]. In the United States, 30 million tonnes of MSW generated about 14 billion kWh of electricity in 2016 [2]. The energy in the MSW is normally recovered using the process of incineration with heat recovery. The heat may be used directly for heating, usually district heating, indirectly for electrical power production, or for both combined heat and power generation. The estimated installed global capacity of power generation using MSW is about 3 GW, and about half of it is in Europe [8].

9.8.5 Commercial and Industrial Wastes

Commercial and industrial (C&I) wastes of organic origin come from a wide range of commercial activities, such as manufacturing, agricultural, and services industries. In 2016, the United Kingdom generated about 40.1 million tonnes of C&I wastes, of which 31.7 million tonnes were generated in England, and for 2017, this figure (i.e., for England) climbed up to 38 million tonnes. C&I wastes from the food industry, such as fats either in the form of used cooking oils or unwanted fatty tissue removed during meat processing, are potentially suitable either for direct combustion or conversion to biodiesel. Another important example of C&I wastes is worn tires; for example in the United Kingdom, it is estimated that 50 million tires are discarded each year with an energy content of 15 GJ per tonne, representing a valuable source of energy [8].

Example 9.5

It is claimed that if 10 million square kilometers of world land area is devoted to growing energy crops, it could contribute to annual bioenergy of about 250 EJ, which is approximately equivalent to 50% of the current global primary energy consumption.

a. Estimate the average annual bioenergy output per hectare that would be required to provide 250 EJ per year from the land area suggested above.

b. Given that the average energy density for maize and sugarcane is 17 GJ per tonne and their average annual yields are 35 and 15 tonnes per hectare respectively, calculate the annual energy yield of each in GJ per hectare. Compare your results with the claim made at the start of the question.

Solution:

10 million square kilometres is

$$1.0 \times 10^7 \ \text{km}^2 = 1.0 \times 10^7 \times 10^6 \ \text{m}^2 = 1 \times 10^{13} \ \text{m}^2$$

One hectare (ha) is 10,000 m²; therefore, in hectares (ha) this is

$$\frac{1 \times 10^{13} \ \text{m}^2}{10000} = 1 \times 10^9 \ \text{(ha)}$$

Therefore, the average bioenergy output per hectare per year is

$$\frac{250 \times 10^{18} \ \text{J}}{1 \times 10^9 \ \text{(ha)}} = 2.50 \times 10^{11} \ \text{J/ha.yr}$$

In GJ this is

$$\frac{2.5 \times 10^{11}}{10^9} = 250 \ \text{GJ/ha.yr}$$

For sugarcane the average annual yield is 35 tonnes per hectare; therefore the annual energy yield is

$$35 \times 17 = 595 \ \text{GJ}$$

This is nearly 2.4 times the figure of 250 GJ in part (a) above. For maize the annual yield is

$$15 \times 17 = 255 \ \text{GJ/ha}$$

This is just 2% more than the claimed value of 250 GJ from part (a).

Example 9.6

Repeat Example 9.5 above using a grade of miscanthus with an annual yield of 12 tonnes per hectare and energy density of 10 GJ per tonne. Explain your results.

9.9 Advantages of Biomass

As the race continues for finding an alternative for fossil fuels, interest, and investment in biomass has been accelerating due to many of its advantages, which we discuss next.

9.9.1 Renewable

If well-managed, many sources of biomass, such as trees, maize, and sugarcane, can be readily replenished after use, which makes biomass truly renewable sources of energy. With replanting after responsible harvesting, plants can be maintained as a lasting source of energy.

9.9.2 Carbon-Neutral

When plants are used as fuel, only carbon dioxide released into the atmosphere is what was taken in during their life cycle. This carbon will be reclaimed from the atmosphere when these used plants are replenished. That is, this process of burning and replanting new plants is carbon-neutral and this is an exceptional feature of biomass.

9.9.3 Versatile

Biomass sources come in many different physical forms and can be converted into a variety of different forms of fuels. For example, they can be converted to liquid biofuels to power different types of machineries, or gaseous fuels for cooking, heating, and electricity generation.

9.9.4 Reduction of Waste

Recycling of household refuse, human and animal wastes, and commercial and industrial wastes provide, useful energy and also helps keep the environment clean and reduces the emissions of harmful greenhouse gases into the atmosphere.

9.9.5 Local

The local aspect is an important feature of biomass, which means that energy can be generated locally and cheaply. For example, a farmer can readily set up a small plant that uses the manure of cattle to produce methane gas, which can then be used for heating and cooking.

9.9.6   Abundant

Biomass sources in one form or another can be found or can be made available almost anywhere on Earth. This is because many biomass sources are integral part of the natural life cycle of our planet. However, to make sure that this remains sustainable, it must be carefully managed.

9.10   Disadvantages of Biomass

There are some claimed disadvantages to using biomass for energy production, which we discuss next.

9.10.1   Not Completely Clean

Even though biomass is a mostly carbon-neutral source of energy, the burning of wood still produces other emissions and smoke that pollutes the local environment, although the effects are much less than those produced by burning fossil fuels. However, compared to other renewables, such as wind and tidal, these effects are more evident.

9.10.2   Risk of Deforestation

Biomass fuels are sustainable sources of energy only if their resources are well-managed and maintained, otherwise the result will be widespread deforestation. Moreover, there is the argument that cutting down trees deprives wildlife of their natural habitat, which may lead to their extinction.

9.10.3   Requires Large Space

Planting crops for energy requires relatively large amounts of space, which may not always be available; for example, in built-up areas. This may mean having to plant biomass far away from the intended location of use and or processing, which results in added costs of transport. This makes them less favorable compared to other renewables such as solar energy. There is also the debate about using land for energy rather than for food production, which may lead to increased food prices.

9.10.4   Inefficient

Biofuels are not as efficient as fossil fuels. Fossil fuels such as diesel and petroleum are usually added to their biofuels counterparts to improve efficiency. This reduces the effectiveness of biofuels in cutting down dependence on fossil fuels.

9.10.5 Requires Water

One clear disadvantage of biomass is the large amount of water energy crops need. This can be costly especially if the source of water is remote and/or scarce. Moreover, this may result in water supply becoming less available for wildlife and/or humans. There is also the debate that water may also be used as a source of renewable energy, which is cleaner than biomass.

9.11 Processing of Biomass

Biomass fuel resources exit in a variety of physical forms (e.g., plants wood offcuts, liquids, and solids). Although in some cases they can be burned directly to release their stored energy; for example burning wood wastes in households, in most cases and certainly at large scales, biomass fuels require some processing before they can be used. This processing might be necessary for many reasons; for example, to make them more user-environment-friendly, easier for transportation, or make it easier to release their energy. The processing of biomass fuels can involve one or more of physical, thermochemical, or biochemical processes [8]. Examples of physical processing are the cutting and chipping of wood and the pressing of oily seeds. Cutting wood into smaller pieces makes it easier to transport and to use. Pressing of oilseeds is necessary to extract their oil, which is the end-user product.

Thermochemical processing involves the use of heat and possibly the use of some chemical reagents to convert biomass into more useful forms. The output from such processes may be heat, intermediate gaseous, or liquid fuels. An example of such a process is the combustion of solid biomass fuels in special ovens or stoves to provide heat. Another example is gasification in which a gaseous biofuel fuel is produced from solid biomass, such as woodchips under relatively high temperature in a carefully controlled environment that typically contains oxygen. The gas produced is called synthesis gas or syngas, which consists primarily of hydrogen, carbon monoxide, and carbon dioxide. Syngas is then cleaned refined and liquefied for use as a renewable energy fuel [8].

Biochemical processes almost always involve physical and chemical processing in addition to the fundamental biological process. These processes depend on microorganisms to convert biomass fuel into more useful and useable form of biofuel. A common example of biochemical processing is anaerobic digestion. This process uses bacteria to break down organic material into various organic acids, which after further decomposition, produces biogas that consists primarily of methane and carbon dioxide. Examples of the feedstock used are dung, sewage, food wastes and or refuse, and agricultural residue. The process takes place in a digester where digestion can be either wet or dry. In a wet digester, the feedstock (e.g., dung) is converted to a slurry by mixing it with water. This requires a relatively large amount of water, which is clearly a disadvantage. In a dry digestion process, the moisture content in the digester is reduced, which means more energy input is required to mix the material; however, it alleviates the need for large quantities of water. Another example of biochemical processes is the process of fermentation. This is also an anaerobic process that uses yeast as the microorganism and sugar as the biomass fuel to produce ethanol.

For a detailed and well-cited treatment on the processing of biomass, the reader is referred to [8].

9.12   Summary

Biomass is organic material that comes from plants and animals in a variety of physical forms and varying levels of energy contents. Biomass is a renewable source of energy that stores significant amounts of solar energy. Biomass plants absorb the sun's energy during the process of photosynthesis. When biomass is burned, the chemical energy stored in it is released as heat. Some biomass may be burned directly to generate heat and/or steam to turn turbines to generate electricity or converted to liquid biogas that can be used as fuel for generating mechanical energy.

Examples of biomass:

- Woody plants and wood processing wastes;
- Agricultural crops and waste materials;
- MSW and household refuse;
- Animal manure and human sewage.

Biomass is abundant, environmentally friendly, and sustainable if well-managed and has a good potential as a major source of renewable energy to reduce dependence on fossil fuels. However, it can take up large areas of land and requires large amounts of water and if not well-managed can lead to deforestation.

9.13   Problems

P9.1   The chemical equation showing the combustion of ethanol is

$$C_2H_5OH + 3O_2 \rightarrow 2CO_2 + 3H_2O + \text{Heat energy}$$

Calculate the amount of carbon dioxide generated when one tonne of ethanol is burned.

P9.2 In a given location, the annual yield per hectare of ovendried wood (i.e., with 0% moisture content) is 20 tonnes. Given that the energy density of ovendried wood is 18 GJ per tonne, calculate the amount of annual energy in joules and in kWh that can be generated by an area of one million square kilometers when used to produce oven-dry wood.

P9.3 Use data from Table 9.1 to answer this question. The volume of one tonne of a certain grade of coal is 0.53 m^3. Calculate the volume of each of the following: stacked newspapers, domestic refuse, and maize, biofuels that would be required to supply the same amount of energy as one tonne of this coal, and explain your answers.

P9.4 A land of 10 square kilometers is to be dedicated to biofuel energy crops. There is a choice of miscanthus or maize. Given that the average energy content for miscanthus and maize is 13 and 15 GJ per tonne respectively and that their average annual yields are 12 and 19 tonnes per hectare, respectively, calculate the average annual bioenergy output in kWh for each choice of biofuel.

P9.5 An area of 20,000 hectares is used to grow miscanthus to provide district heating. If miscanthus has a yield of 12 tonnes per hectare per year and an energy content of 18 GJ per tonne:

a. Calculate the annual energy supply represented by this area.

b. Assuming a household needs on average 3,000 kWh a year for heating and that 30% of the energy produced by the miscanthus is lost during processing and delivery, how many households could be heated by this area?

P9.6 A 40-MW straw-fired power plant uses 250,000 tonnes of straw per year and its average annual electricity output is 300 GWh. Given that the energy density of straw is 15 GJ per tonne, calculate the average annual efficiency of straw to electricity conversion and the average capacity factor of the plant. What are the main disadvantages of using straw as a fuel for a power station?

P9.7 The council of a small town of 18,000 households is considering the possibility of an energy from waste (EfW) power generation plant to mitigate the cost of landfills for its annual 20,000 tonnes of domestic refuse. Assuming that the energy density of household refuse is 10 GJ per tonne:

a. Calculate the number of annual kilowatt-hours the proposed plant would generate, using all the local household refuse and running at a waste-to-energy conversion efficiency of 30%.

b. Using the result of (a) above, determine a suitable rating in kW for the generator assuming the plant will be running with a capacity factor of 82%;

c. Given that the average household consumption of electricity is 3,200 kWh per year, what fraction of the total load demand would be provided by the plant?

P9.8   Explain why biomass is, in principle, a solar source of energy.

P9.9   Explain what is meant by the following terms: bioenergy, primary and secondary biomass, and carbon-neutral energy.

P9.10   Explain what makes biomass a source of energy and what makes it a sustainable source of energy.

P9.11   Write a concise essay why it is important to develop large-scale systems to produce biofuels for motor vehicles.

P9.12   Write a short essay of about 500 words in which you give your own critical views on the question of using land to produce energy crops.

P9.13   Discuss a few ways to reduce your own household refuse.

P9.14   Discuss two advantages and two disadvantages of biomass as source of energy.

P9.15   What are the shortcomings of using straw as the main fuel for a power generation plant?

P9.16   Recently, there has been a growing global interest in the production of liquid biofuels from purpose-grown energy crops. Using appropriate literature, write an appropriately referenced paper of about 800 to 1,000 words that includes the following:

1. A description of two types of primary energy source plants outlining briefly the methods used to extract the biofuels from them.

2. A discussion of possible benefits of a major shift from fossil-based liquid fuels to liquid biofuels and the main obstacles, and critically evaluate the debate for and against the use of land for energy rather than for food.

3. An assessment of the current use and future potential of liquid biofuel.

P9.17   Use the literature to write an appropriately referenced essay of about 1,000 words assessing and critically reviewing the current and future potential of using liquid biofuels in the transportation industry.

References

[1]   International Energy Agency, "Bioenergy and Biofuels," https://www.iea.org/topics/renewables/bioenergy/.

[2]   U.S. Energy Information and Administration, "Biomass Explained," U.S. Energy Information and Administration, Washington, August 22, 2018.

[3]   World Bioenergy Association, "Global Bioenergy Statistics 2018," WBA, Stockholm, 2018.

[4]   Slade, R., R. Saunders, Robert Gross, and A. Bauen, *Energy from Biomass: The Size of the Global Resource,* Imperial College Centre for Energy Policy and Technology and UK Energy Research Centre, London 2011.

[5]   Haberl, H., K. H. Erb, F. Krausmann, et al., "Quantifying and Mapping the Human Appropriation of Net Primary Production in Earth's Terrestrial Ecosystems," *Proceedings of the National Academy of Sciences of the USA,* Vol. 104, No. 31, 2007, pp. 1294–1297.

[6]   Thranab, D., T. Seidenberger, J. Zeddies, and R. Offermann, "Global Biomass Potentials—Resources, Drivers and Scenario Results," *Energy for Sustainable Development,* Vol. 14, No. 3, 2010, pp. 200–205.

[7]   Shafiee, S., and E. Topal, "When Will Fossil Fuel Reserves Be Diminished?" *Energy Policy,* Vol. 37, No. 1, 2009, pp. 181–189.

[8]   Boyle, G. (ed.), *Renewable Energy: Power for a Sustainable Future,* Oxford, UK: Oxford University Press/Open University, 2012.

[9]   Forest Research, "Typical Calorific Values of Fuels," Forest Research, 2019, https://www.forestresearch.gov.uk/tools-and-resources/biomass-energy-resources/reference-biomass/facts-figures/typical-calorific-values-of-fuels/.

[10] Matsuoka, S., A. J. Kennedy, E. D. S. Gustavo, A. L. Tomazela, and L. C. S. Rubio, "Energy Cane: Its Concept, Development, Characteristics, and Prospects," *Hindawi Advances in Botany,* Vol. 2014, 2014, p. 13.

[11] Macedo, I. C., J. E. A. Seabra, and J. E. A. R. Silva, "Greenhouse Gases Emissions in the Production and Use of Ethanol from Sugarcane in Brazil," *Biomass and Bioenergy,* Vol. 32, No. 7, 2008, pp. 582–595.

[12] Waclawovsky, A., P. M. Sato, C. G. Lembke, and P. H. S. G. M. Moore, "Sugarcane for Bioenergy Production: An Assessment of Yield and Regulation of Sucrose Content," *Plant Biotechnology Journal,* Vol. 8, 2010, pp. 263–276.

[13] El-Bassam, N., *Handbook of Bioenergy Crops,* London: Earthscan, 2010.

[14] U.S. Environmental Protection Agency, "Municipal Solid Waste," March 29, 2016, https://archive.epa.gov/epawaste/nonhaz/municipal/web/html/.

[15] Department for the Environment Food & Rural Affairs, "UK Statistics on Waste," Department for the Environment Food & Rural Affairs, London, March 7, 2019.

[16] Department for Environment Food & Rural Affairs, "Energy from Waste: A Guide to the Debate," Department for Environment Food & Rural Affairs, London, 2014.

10

Costing a Renewable Energy Project

10.1 Learning Outcomes

After actively engaging with the material in this chapter, you should be able to

1. Explain the factors affecting the cost of generating electricity from a renewable energy plant.
2. Define the terms: capacity factor, efficiency, levelized cost, and payback time.
3. Calculate the payback time for a renewable energy plant.
4. Calculate the cost per unit of electricity from a given renewable energy plant.

10.2 Overview

The chapter presents two simple methods for the financial appraisal of a renewable energy project. The first method is based on calculating the payback time of a project by comparing the cost of the generated electricity against the price of a competing fossil-fueled plant. The second and more realistic method is based on the discounted cash flow method to calculate the cost per unit of generated electricity. Only renewable energy plants that generate electricity are considered.

10.3   Introduction

Renewable energy has entered a definitive trend of falling costs and accelerated technological advances coupled with rising efficiencies. An exception is certain biofuels that have to be purchased, such as wood. Between 2010 and 2016, the cost of electricity generated from solar photovoltaics has fallen by almost 70% and that from onshore wind plants fell by 18% [1]. Continued falling prices and advances in renewable energy technologies are bringing prices of electricity from renewables in line with many competing fossil-fueled power plants. In order to evaluate the economic viability of any renewable energy project, or any project for that matter, it is imperative that some financial appraisals are performed so that a decision to reject or accept a project can be taken. Indeed, it is the objective of this chapter to provide an introductory background on how to assess the economic viability of a proposed renewable energy power plant.

In an ideal world, a power plant should be able to produce electricity 24 hours a day, 7 days a week for the whole year. However, in the real world, this is not possible due to many reasons. For example, a power plant may have to be shut down for regular maintenance and servicing or replacement of faulty or aging parts. In addition, many renewable energy plants have an added limitation on their availability to generate electricity because their output energy depends on weather conditions, which are continuously changing. For example, a solar PV plant cannot produce electricity in the absence of the sun. Similarly, a wind turbine cannot produce electricity without wind. Further, even if sun or wind energy is available, it may not be sufficient to enable a PV or a wind plant to run at its full rated capacity.

10.4   Rated Capacity and Capacity Factor

The rated capacity, also known as the installed capacity, of an electricity generating plant defines the maximum intended full-load output of the plant. The basic unit for measuring the rated capacity is the kilowatt (kW) or any of its multiples; for example, the megawatt (MW) and gigawatt (GW).

The amount of electrical energy a plant can deliver is expressed in kilowatts, megawatts, or gigawatts. There are 8,760 hours in one year (=24 hours × 365 days); hence, if the power rating of a power plant is 1 kW, then the amount of energy it can deliver during a whole year if it runs nonstop, at full capacity, will be 1 kW×8760 = 8760 kWh. However, if the plant runs only for, say, 3,000 hours per year on average at full capacity, then the amount of energy it will deliver is 1 kW×3000 = 3000 kWh. A figure of merit for a power plant is its CF, also known as the load factor, which is defined as the ratio of the actual amount of electricity generated in one year to the theoretical (or maximum) amount of energy a plant can deliver if it runs constantly throughout the year at full rated capacity. The capacity factor may be expressed as a decimal fraction or a percentage as

$$CF = \frac{\text{Actual amount of electricity generated in one year}}{\text{Rated capacity} \times 365 \text{ (days)} \times 24 \text{ (hours)}} \times 100\% \quad (10.1)$$

Using the capacity factor, the actual amount of electricity produced per year can thus be expressed as

$$E \text{ (kWh)} = \text{Rated capacity (kW)} \times \text{Capacity Factor} \times 365 \text{ (days)} \times 24 \text{ (hrs)} \quad (10.2)$$

Note that the capacity factor is a measure of a plant's availability during a year for generating electricity. For fossil-fueled and even some renewable energy plants, such as tidal, the CF factor can be as high as 95%. However, for photovoltaic plants in cloudy climates, the CF can be as low as 10% and for wind turbine plants, it ranges from about 20% to 40% depending on climate conditions [2].

10.5   Types of Costs of a Power Plant

In general, the cost of electricity derived from any power plant may be categorized into four main types of costs [3]:

1. Initial capital costs;
2. Fuel costs;
3. Operation and maintenance (O&M) costs;
4. Decommissioning costs.

Fuel costs are the cost of the primary fuel (e.g., coal in a coal-fired plant) used to generate electricity and may include the cost of fuel transport as well as disposal, such as in the case of nuclear plants. The cost of fuel for most renewable energy plants may be reasonably considered to be zero; an exception is a biomass plant that uses biomass fuel that must be purchased, such as wood, sugarcane, and maize. The initial costs include elements like the costs of land and any required infrastructure in addition to the cost of the initial construction of a plant. Construction costs of nuclear and large-scale renewable energy plants, such as tidal, are typically larger than gas-fired plants, but fuel costs are either very low or zero. However, the cost of gas for a gas-fired plant can be as much as 70% of the cost of generated electricity. Decommissioning costs of a nuclear plant are typically very high. In order to make a comparison between the costs of electricity from different plants, a simple method called the levelized cost of electricity (LCOE) is used here. In this method, the cost of energy is levelized over the lifetime of the plant and is expressed in pence per kilowatt-hour (p/kWh). The main advantage of the LCOE method is the fact that it enables us to make a comparison of the cost of electricity from different technologies, for example, wind, tidal, and coal, of different lifespans, rated capacity, capital cost, and so on. This allows us to make statements such as the cost of electricity from a coal-fired plant is 6 p/kWh and from a wind plant is 4 p/kWh.

10.6   Factors Affecting the Cost of Electricity

The major factors affecting the costs of electricity from a generating plant include the rated capacity and capacity factor, operation and maintenance costs, fuel costs and plant efficiency, and capital cost and the life expectancy of the plant.

10.6.1   Rated Capacity and Capacity Factor

The rated capacity of an electricity-generating plant and its capacity factor are the prime factors that determine the annual amount of electricity that can be generated by a given plant.

10.6.2   Fuel Costs and Efficiency

The primary fuel for many renewable energy plants, such as solar PV, tidal, and wind, is free. However, for a biomass-generating plant the cost of fuel can be comparable to that of fossil fuels. A major contributing factor to the overall cost of fuel in any plant is the energy conversion efficiency of the plant defined as

$$\text{efficiency} = \frac{\text{Energy output}}{\text{Enenrgy input}} \times 100\% \tag{10.3}$$

The efficiency can range from about 10% for photovoltaics plants to about 40% for modern fossil-fueled plants. The fuel cost per kilowatt of energy is expressed as

$$\text{Fuel cost per kWh generated} = \frac{\text{Cost of enenrgy input}}{\text{Efficiency}} \tag{10.4}$$

The contribution to the final electricity cost is usually expressed in pence per kilowatt or pence per megwatt.

10.6.3   O&M Costs

The O&M costs include costs such as those of regular maintenance and safety inspection, repairs and replacements of faulty equipment, insurance, and waste disposal. In addition, these may include costs of access to the plant, which may be significant, for example, for offshore wind plants.

10.6.4   Capital Costs and Plant Lifetime

For most renewable energy plants, the capital costs are the most significant. These are expressed in pounds per kilowatt of rated capacity. This allows us to make statements such as the cost of an offshore wind plant is £3,000/kW and for onshore wind turbine plant it is £1,000/kW. Since the capital costs of a renewable energy plant are relatively high, the life expectancy of a plant is a major factor in assessing its economic viability. Typically, wind turbines and photovoltaic modules have life expectancies of about 20 to 30 years and a hydroelectric plant may have a life expectancy in excess of 60 years.

10.7   Calculating Costs

We will now consider two simple methods for assessing the economic viability of a renewable energy project. The first is a simple method based on the payback time calculation, while the other is based on levelized costing.

10.7.1   Payback Time Method

One simple method of assessing the economic viability of a project is to calculate the time it takes the project to recover the initial capital outlay, known as the payback time, which is normally expressed in years. The inputs to this method are the capital cost, the price of the output energy, and the price of a competing energy source, which is typically a fossil source of energy. The latter is then compared with the price of the output energy to calculate the payback time. The operation and maintenance and any finance costs are not taken into account, which artificially reduces the calculated payback time.

Example 10.1

A proposed onshore wind energy plant is rated at 2 MW. It has a capital cost of £1,800/kW and an estimated O&M costs of £35,000 per year. It has a life expectancy of 25 years and a CF of 25%. Assuming a competing fossil fuel electricity price of 15 pence per kilowatt, estimate the payback time of the plant.

$$\text{Total capital cost} = \text{Rated capacity (kWh)} \times \text{Capital cost per kW}$$
$$\therefore \text{Total capital cost} = (2 \text{ MW} \times 1000) \times £1800 = £3\,600\,000$$

The annual electricity generated, E(kWh/y) is

$$E \text{ (kWh/y)} = \text{Rating (kW)} \times \text{CF} \times 365 \text{ (days)} \times 24 \text{ (hrs)}$$

That is

$$E = 2000 \text{ kWh} \times 0.25 \times 24 \times 365 = 4.380 \times 10^6 \quad \text{kWh/y}$$

Using the competing fossil price of 15 pence per kWh, we can calculate the price of the electricity generates as

$$\text{Value of annual electricity generated} = 4.380 \times 10^6 \times £0.15 = £657000 \text{ /y}$$

Hence, the payback time is calculated as

$$\text{Payback time} = \text{Total capital cost} / \text{value of annual output}$$
$$\text{Payback time} = £3600000 / £657000$$
$$= 5.5 \text{ years}$$

Note that this simple payback method ignores the operation and maintenance costs of £35,000/year, which if included would elevate the payback time significantly.

10.7.2   Annual Levelized Cost of Electricity

The LCOE represents the average return per unit of electricity generated by spreading various costs of the plant over its life expectancy. The main inputs required to calculate the LCOE include the initial capital costs, O&M costs, and finance costs (e.g., interests on loans). For renewable plants with no fuel costs, such as solar and wind plants, the LCOE changes in proportion with the capital costs, while for plants with high fuel costs, the LCOE varies considerably with the costs of fuel. The levelized cost can be calculated on annual basis and the capital cost is repaid in equal annual payments over the lifetime of the project. Such a stream of equal payments is known as an annuity. We will now consider two levelized costing methods; the first assumes zero cost of interest and the second uses the method of discounted cash flow to allow for the cost of interest.

In the first annual levelized method, the capital is assumed to be repaid in equal annual payments over the lifetime of the project without incurring any interest charges. The levelized O&M and fuel costs are added to the annual payments. The levelized (average) cost of electricity is given by

$$\text{Payback time} = \text{Total capital cost} / \text{value of annual output}$$
$$\text{Payback time} = \text{£3600000} / \text{£657000}$$
$$= 5.5 \ \textit{years} \tag{10.5}$$

The method is illustrated with the following example.

Example 10.2

For the above wind turbine plant, the cost of capital spread over the 25 years design life of the plant is:

$$\text{Cost of capital spread over 25 years } = \text{£3 600 000} / 25 = \text{£144000}$$

Adding the O&M cost, the total overall annual cost is

$$\text{£144000} + \text{£35000} = \text{£179000}$$

Therefore, the overall cost of electricity per unit is

$$\text{£179000} / (4.380 \times 10^6) \ (\text{kWh}) = \text{£0.041 per kWh} \quad \text{or} \quad 4.1 \ \text{p/kWh}$$

In the above method, it was assumed that the total capital cost will be repaid in equal amounts without any interest charges. However, in practice, the value of money changes with time. If you invest a sum of money, say V_P in a saving bank account with an interest rate of say r%, then after n years, the future value of your investment will be

$$V_n = V_P(1+r)^n \tag{10.6}$$

For example, if you invest £100 today, after one year this is worth $100(1+0.1) = £110$ and after two years this is worth £121. Stated in other words, the present value of £121 in two years' time at a rate of 10% is only £100. This is called cash discounting because we are expressing today's value of the future value of a sum of money in terms of a smaller sum. This leads to a method of project financial assessment called discounting cash flow (DCF). The future amount of money V_n can be discounted by rearranging the above equation:

$$V_p = \frac{V_n}{(1+r)^n} \tag{10.7}$$

If a capital sum, say V_p, is to be repaid as an annuity (i.e., annual equal amounts over the duration n years of a loan) using a discount rate r, the annuity (A) is calculated using the following formula:

$$A = \frac{rV_p}{1-(1+r)^{-n}} \tag{10.8}$$

Example 10.3

Now let us revisit the above example and assume a discount rate of 10% to calculate the cost of the project using the levelized costing with DCF method.

The first step is to find the value of the annuity of the capital cost as

$$A = \frac{rV_p}{1-(1+r)^{-n}}$$

$$A = \frac{(0.1)(3600000)}{1-(1+0.1)^{-25}} = £396605$$

The total annual costs are the sum of the capital cost and the O&M cost; that is

Total annual cost per kWh $= £396605 + £35000 = £431605$

Therefore, the overall levelized cost per kilowatt of generated electricity is

Overall cost per kWh $= £431605 / 4380000 \text{ kWh} = 9.85 \text{ p/kWh}$

Thus, using a discount rate of 10% and a design lifetime of 25 years, we find that the cost is 9.85 pence per kilowatt, which is much higher than the 4.1 pence per kilowatt predicted by the simple levelized method without discounting.

Example 10.4

A proposed biomass wood-fired power plant with a rating of 10 MW has a capital cost of £2,000 per kilowatt and is expected to run for 8,000 hours, on average, per year at full power. The plant has a life expectancy of 30 years and an estimated overall electrical generation efficiency of 30%. O&M costs have been estimated at 2 pence per kilowatt of electricity generated. It is assumed that wood will be available at £5 per GJ over the lifetime of the plant. Calculate

a. The capacity factor and amount of electrical energy generated per year.

b. The total capital cost of the plant.

c. The total cost per kWh of electricity generated using the DCF method using a discount rate of 10%.

Solution:

(a) The capacity factor is

$$CF = \frac{8000 \text{ hours}}{24 \times 365} \times 100\% = 91\%$$

The amount of electricity generated per year is

$$E = 8000 \text{ hours} \times 10 \text{ MW} = 80\,000 \text{ MWh}$$

The cost of fuel is £5 per GJ

$$1 \, GJ = \frac{10^9}{1000 \times 60 \times 60} = 277.7 \text{ kWh}$$

Therefore, the fuel cost per kilowatt is

$$\frac{£5 \times 100}{277.7 \text{ kWh}} = 1.8 \text{ p /kWh}$$

(b) The total capital cost

$$\text{The total capital cost} = \text{Rated capacity} \times \text{Cost per unit}$$

hence

$$\text{The total capital cost} = 10\,000 \text{ kW} \times £2000/\text{kW} = £20 \text{ million}$$

Therefore, for a borrowed sum of £20 million the annual repayments will be

$$A = \frac{(0.1)(20\,000000)}{1-(1+0.1)^{-30}} = £2.12 \text{ million}$$

Annuitized capital cost per kilowatt of electricity produced is

$$£2.12 \times 10^6 / 80\,000 \text{ MWh} = £26.5/\text{MWh} \quad \text{or} \quad 2.65 \text{ p /kWh}$$

The fuel cost per kilowatt of electricity generated is

$$\frac{1.8 \text{ p /kWh}}{\text{efficiency}} = \frac{1.8 \text{ p /kWh}}{0.30} = 6 \text{ p/kWh}$$

Operation and maintenance costs equals 2 p/kWh

$$\text{Total cost per kWh} = 6 + 2 + 2.65 = 10.65 \text{ p/kWh}$$

Note that in this example the fuel costs account for more than 50% of the final electricity cost.

10.8   Problems

P10.1   A fluorescent lamp costs £5 and uses only 20% of the electricity of a 100W incandescent bulb of equivalent light that costs only 40 pence. In addition, a fluorescent lamp will last approximately 8,000 hours against only 1,000 hours for an incandescent bulb. Assuming electricity costs 15p per kilowatt, calculate the cost of using each type of lamp for a period of 8,000 hours.

P10.2   A new heating system costs £2,500 and is expected to save £200 on the annual electricity cost. Using simple payback calculations, how many years will it take to recover the cost of the system?

P10.3   A proposed wind turbine power plant is rated at 100 MW and has a capital cost of £1,500 per kilowatt. It is expected to run for 2,500 hours per year at full power and has a design lifetime of 25 years. Its overall electrical generation efficiency is 40%. Operating and main-tenance costs have been estimated at 1.2 p per kilowatt of electricity generated. Calculate the cost of the generated electricity in pence per kilowatt ignoring any capital interest charges.

P10.4 Repeat the previous exercise using discounting assuming a discount rate of 10% over 25 years.

P10.5 Use the simple discounted cash flow method to calculate the cost of electricity from a proposed 1 MW wind turbine plant which has a capital cost of £1,500 per kilowatt and a life expectancy of 25 years. Assume that its capacity factor is 30% and its O&M costs are 1.5 pence per kilowatt and a discount rate of 8%.

P10.6 A proposed biomass wood-fueled 50 MW power plant has a capital cost of £2,200 per kilowatt and a capacity factor of 80% and has a design lifetime of 30 years. Its overall electrical generation efficiency is 35%. O&M costs have been estimated at 1.5 p per kilowatt of electricity generated. It is assumed that wood will be available at £8 per GJ over the lifetime of the plant. Calculate

a. The amount of electrical energy generated per year in megawatts;

b. The total capital cost of the plant;

c. The total annuitized capital cost per kilowatts of electricity generated using the DCF method with a discount rate of 10%.

P10.7 The council of a small town of 20,000 households is considering the possibility of an energy from waste (EfW) power generation plant to mitigate the cost of landfills for its annual 25,000 tonnes of domestic refuse. Given that the energy density of household refuse is 9 GJ per tonne

a. Calculate the number of annual kilowatt-hours the proposed plant would generate, using all the local household refuse and running at a waste-to-energy conversion efficiency of 30%.

b. Using the result of (a) above, determine a suitable rating in kilowatts for the generator assuming the plant will be running with a capacity factor of 80%.

c. Given that the average household consumption of electricity is 3,500 kWh per year, what fraction of the total load demand would be provided by the plant?

d. Assuming the capital cost is £3,500 per kilowatt-hours. Determine the total capital cost of the plant.

e. The council decides to borrow the total capital cost at a rate of 10% to be repaid over 15 years. Calculate the annual payment needed.

f. The operation and maintenance costs are expected to be £200,000 per year. However, it is expected that a saving of £30 per tonne will be made by avoiding landfill fees. Calculate the total annual cost of the plant.

g. Calculate the cost of kilowatt-hours generated by the plant.

References

[1]   I. R. E. Agency, "International Power Generation Costs 2017," https://www.irena.org/publications/2018/Jan/Renewable-power-generation-costs-in-2017.

[2]   Boyle, G. (ed.), *Renewable Energy Power for a Sustainable Future*, Oxford, UK: Oxford University Press with The Open University, 2012.

[3]   Committee on Climate Change, "Costs of Low-Carbon Generation Technologies," 2010, https://www.theccc.org.uk/archive/aws/Renewables%20Review/MML%20final%20report%20for%20CCC%209%20may%202011.pdf.

About the Author

Dr. Nader Anani received his B.Sc. (Hons) degree in electrical and electronic engineering from Newcastle University (UK) in 1984. He lectured at the College of Science and Technology in Jerusalem from 1984–1986. In 1987 he obtained his M.Sc. degree in microelectronics from the University of Manchester (UK). In December 1987, he took up the position of research assistant at the Manchester Metropolitan University, working on harmonics elimination and voltage control in voltage-source inverters for traction applications. In March 1990, he took up the position of Research Fellow at the University of Sheffield (UK), working on the design of magnetizing fixtures for high-energy rare-earth permanent magnets. From 1992–1993, he studied for his postgraduate diploma in education at the University of Huddersfield (UK). After a short period as a lecturer at Bridgwater College (UK), he joined the Manchester Metropolitan University (MMU) as a senior lecturer in electrical and electronic engineering in February 1994. In 2010 he started studying for his Ph.D., which he completed in 2013. From 2014–2017, Dr. Anani was a senior lecturer in electrical systems at Sheffield Hallam University. In June 2017, he moved to the University of Chichester as a founding head of electronics and electrical engineering. In September 2019, Dr. Anani took the position of a reader (associate professor) at the University of Wolverhampton (UK). His research interests are in renewable energy and power electronics, and he has supervised and examined over 20 Ph.D candidates. Dr. Anani is a senior member of the IEEE and a chartered UK engineer.

Index

For further information on these and other Artech House titles, including previously considered out-of-print books now available through our In-Print-Forever® (IPF®) program, contact:

Artech House	Artech House
685 Canton Street	16 Sussex Street
Norwood, MA 02062	London SW1V 4RW UK
Phone: 781-769-9750	Phone: +44 (0)20 7596-8750
Fax: 781-769-6334	Fax: +44 (0)20 7630-0166
e-mail: artech@artechhouse.com	e-mail: artech-uk@artechhouse.com

Find us on the World Wide Web at: www.artechhouse.com